Climate-Smart Agriculture

Samiha Ouda · Abd El-Hafeez Zohry

Climate-Smart Agriculture

Reducing Food Insecurity

Springer

Samiha Ouda
Water Requirements and Field Irrigation Research Department
Soils, Water and Environment Research Institute; Agricultural Research Center
Giza, Egypt

Abd El-Hafeez Zohry
Crops Intesefication Research Department
Field Crops Research Institute; Agricultural Research Center
Giza, Egypt

ISBN 978-3-030-93110-0 ISBN 978-3-030-93111-7 (eBook)
https://doi.org/10.1007/978-3-030-93111-7

This Springer imprint is published by the registered company Springer Nature Switzerland AG
The registered company address is: Gewerbestrasse 11, 6330 Cham, Switzerland

To Our Parents
May Allah Have Mercy on Their Souls

To Our Children and Grandchildren
May Allah Bless Them with Bright Future and Happiness

Preface

In 2015, the UN General Assembly released what is known as 'Transforming Our World: The 2030 Agenda for Sustainable Development'. The Agenda has 17 Sustainable Development Goals (SDGs), 169 targets demonstrate the scale, and ambition of this new universal Agenda to be a plan of action for people, planet, and prosperity. The agenda integrated and attain the balance between the three dimensions of sustainable development: economic, social, and environmental.

Among the 17 SDGs, Goal 2 seeks to end hunger and ensure access by all people, in particular, the poor and people in vulnerable situations, to safe nutritious and sufficient food all year round by the year 2030. Target 4 of that goal seeks to ensure sustainable food production systems and implement resilient agricultural practices that increase productivity and production. It ensure helping maintain ecosystems that strengthen capacity for adaptation to climate change, extreme weather, drought, flooding, and other disasters, which progressively improve land and soil quality. However, more severe and frequent extreme weather events are causing high uncertainty in accomplishing this goal.

The FAO organization produced the concept of climate-smart agriculture to achieve the above target, with emphasis on the effect of climate change. The concept of climate-smart agriculture has fascinated us with its collective practices that could attain better welfare to rural communities in developing countries. Thus, we decided to write this book to elaborate on the concept of climate-smart agriculture and classify some of the practices done in Egypt according to its pillars. Furthermore, we suggested other management practices that can be easily done in a developing countries to improve food security conditions.

In Part I of the book, we will discuss the first pillar of climate-smart agriculture, namely sustainably increases agricultural production. This part consists of four chapters to cover the related subjects to the sustainable use of natural resources to increase food production. The Chap. 1 will discuss the concept of water-smart practices and how it could be used to manage water scarcity in Egypt. The Chap. 2 will present one of the successful practices to deal with water scarcity, namely using fish farms effluents for irrigation in a field and modeling studies. The Chap. 3 will tickle the idea of soil-smart practices through the integrated soil fertility management. Lastly, the Chap. 4 will discuss the integration of crop-smart, water-smart, and soil-smart practices to sustainably increase food security.

Part II of this book will discuss the second pillar of climate-smart agriculture, namely build resilience to climate, which will include three chapters. The Chap. 1 will discuss the most effective influencer in crop production, namely climate extremes and the suggested measures to the stress of climate variability in Egypt. The following chapter will assess the effect of climate variability on wheat productivity in Egypt. The last chapter in this part will review several climate-resilient crops cultivated around the world and in Egypt.

In Part III, the third pillar, namely reduction of the greenhouse gases from the cultivated area will be discussed in one chapter which will review practices that contribute to the reduction of the emission of greenhouse gases.

We hope that this book will assist researchers in other developing countries in their sensible quest to improve their food security situation and attain a better life for their citizens.

Giza, Egypt Samiha Ouda
2021

Introduction

Tremendous pressure, largely exists, on the use of natural resources due to rising population, urbanization, environmental stresses, and climate change. Agriculture production system requires resilient development pathways to augment global food grain production, reduce greenhouse gases emissions from agricultural activity and adapt agriculture to climate change and variability (Lipper et al. 2014). The Food and Agriculture Organization (FAO) of the United Nations has reported that food production needs to be increased by at least 60% to meet the needs of the world's expected population of 9 billion by 2050 (FAO 2014). This will be a great challenge for global agriculture under the condition of climate variability resulted from climate change. Additionally, shifting weather patterns, including rising temperatures, and unpredictable precipitation can have negative effects on crops yield, and agricultural production.

Weather is an important determinant of the productivity of the cultivated crops. Temperature and rainfall affect the development of plants, either alone or by interacting with other factors (IPCC 2007). The World Meteorological Organization (2012) stated that weather reflects short-term conditions of the atmosphere, while climate is the average daily weather for an extended period of time at a certain location over many years (usually at least 30 years). The uncertainty associated with weather patterns is one of the greatest contributors to the gap present between potential and actual yield (Zhang et al. 2013). Lobell et al. (2009) estimated the gap between potential and actual yield by 20% in many cases. Climate variability is often used to describe any kind of change in climate that may be natural or human-induced (Hegerl et al. 2007). Ofori-Sarpong (2011) indicated that rising temperatures, unpredictable rainfall, loss of soil moisture, and increased evaporation and transpiration are the factors of climate variability.

Extreme Weather Events

IPCC (2011) defines the extreme weather events caused by climate variability "as rare as or rarer than the 10th or 90th percentile of a probability density function estimated from observations". Extreme events are expected to cause instability of crops yields and forming high risk to global crop production (Lobell et al. 2013). Research on the effects of extreme weather events on major crops worldwide was employed. Schlenker and Roberts (2009) indicated that extremely high daytime temperatures are destructive and could be lethal to crops. Drought leads to stomata closure and photosynthesis reduction, which negatively affects the yield (Chaves et al. 2010). Rosenzweig et al. (2001) reported that higher temperatures than threshold temperatures may lead to significant reductions in crop yields. Physiological processes of a crop could be affected by extreme weather events causing physical damage (van der Velde et al. 2012). Acceleration in phenological development of crops under warmer conditions is observed, which reduces growing season, reduces net photosynthesis, and consequently reduces yield (Lobell and Ortiz-Monasterio 2007). Other indirect effects could occur, such as influences on soil processes and nutrient dynamics. Moreover, insect, pest, and plant diseases may magnify under exposure to extreme weather events (Motha 2011).

Yield losses higher than 30% in fruit and vegetable crops were found due to drought conditions in 2016 in the Northeastern United States reported (Sweet et al. 2017). It was reported that photosynthesis rates of sweet corn were decreased by 30% when the temperature exceeded 30 °C (Ben-Asher et al. 2008). Moreover, the interaction between high temperature and drought stress reduced both wheat photosynthesis and yield (Shah and Paulsen 2003). Powell and Reinhard (2016) found that high temperature and precipitation events are time specific, where its effect on yields is varied based on growth stages of wheat.

Egyptian agriculture has faced serious risks from climate variability causing significant negative impacts on the yield of major crops and consequently on food security. It was reported that in 1998 growing season, rice plants were exposed to heatwave during flowering and grain filling stages causing a reduction in national rice production (Khalil and Hassanein 2016). Mahmood et al. (2010) indicated that exposing rice plants to heatwave during the flowering growth stage could cause sterility, whereas during the grain filling stage, it reduces grain weight and grain number. They also reported that a decline in grain number is directly proportional to increasing temperature.

Wheat production was also reduced in 2010, as a result of heatwave prevailed in that year by an average of 15% over Egypt (Saleh et al. 2017). Taha (2012) reported losses in wheat productivity in 2010. With respect to maize, Saleh et al. (2017) indicated that maize production was reduced by an average of 11% in 2015 as a result of heatwave that occurred in that year, compared to its production in 2010.

Climate Change Effects

Agricultural production is affected by climate change in direct and indirect pathways. The increase in mean temperatures accelerates directly crop development, the change in seasonal precipitation amounts together with increasing evaporative demand can indirectly lead to more drought stress for crops (IPCC 2012). Martins et al. (2019) indicated that increased temperatures are likely to shorten the crop cycle as a result of acceleration in crop development and phenological stages (Eyshi Rezaei et al. 2018), thus reducing crop production. On the other hand, alteration of precipitation will affect water availability for crops, so that crop yield will be severely affected or even crop failure could occur (Asseng et al. 2013). Thus, global climate change adversely impacts crop production and imposes a wide range of constraints on agricultural systems, especially in water-limited environments (Paymard et al. 2018).

Climate-Smart Agriculture

The deterioration in the natural resources worldwide, as a result of the continual use of the same technologies necessitates the use of sustainable practices to enable to produce food for the growing population and to mitigate the effects of climate change. Climate-smart agriculture could be the approach to be used to avoid the deterioration of natural resources. Climate-smart approach was introduced by FAO (2010) as an integrated approach that aims at sustainably increases agricultural productivity and incomes, build resilience to climate, and reduce emissions of greenhouse gases. Therefore, it put emphasis on the need for changing the prevailing management practices in agriculture including water and soils to ensure its efficient and sustainable use to produce more food, adapt to climate change, and reduce emissions of greenhouse (FAO 2013). Climate-smart agriculture has also been identified as an approach for transforming and reorienting agricultural systems to support food security under the new realities of climate change (Lipper et al. 2014). Climate-smart agriculture has also been known as "triple win", which simultaneously achieves its three objectives (or pillars): food security, mitigation, and adaptation (Fallot 2016). It calls for a set of actions by decision-makers from farm to global level, to enhance the resilience of agricultural systems and livelihoods and reduce the risk of food insecurity in the present, as well as in the future (Lipper et al. 2014). The three pillars of climate-smart practices can be stated as follows:

Sustainably Increases Agricultural Production

Climate-smart practices can attain higher production to feed the growing population and increase food availability and security. It was reported that using sustainable farm management increased crops production, compared to traditional practices (Zohry et al. 2020). Climate-smart practices could include water-smart practices that increase water availability under rain- fed agriculture through water harvesting, storage, and source protection, as well as increase its efficient use under irrigated agriculture. It also includes soil-smart practices that improve soil health through safe farming inputs, reduce soil erosion, and nutrient leaching. Climate-smart agriculture also includes crop-smart practices that promote using intercropping systems, crop rotation, and the use of climate tolerant cultivars (Subedi et al. 2019). Some of these practices can achieve a double positive effect on food production. Intercropping systems, in general, and intercropping with legumes could play a role in water-smart and soil-smart practices. Intercropping contributes to increasing water and lands equivalent ratios, thus saving water and increasing production (Ouda et al. 2018 and Zohry and Ouda 2019), and intercropping with legumes increases soil organic matter and nitrogen contents (Zohry et al. 2017). On the other hand, rotating crops in a crop rotation could also play the same role if intercropping systems with legumes are included, or if cereal crops were rotating with legume crops (Ouda et al. 2018).

Build Resilience to Climate

Build resilience to climate is an important goal in climate-smart agriculture. The primary climate challenges that impact farming are seasonal shifts, ecological disturbances, and extreme weather events, which impact the principle needs of a crop, namely water, sun, and soil. Building resilience is enhancing the ability to adapt to climatic shocks, which can be improved through internal and external interventions (Meuwissen et al. 2019). Building resilience to climate can be identified by three stages, namely reducing exposure, reducing sensitivity, and increasing adaptive capacity (Gitz and Meybeck 2012). Extreme weather events, such as droughts, heat and cold waves, and flooding have serious effects on food production. To cope with it, farmers need to modify farm management practices, such as adjusting planting time, supplementing irrigation in rain-fed agriculture, implementing intercropping systems, adopting conservation agriculture, and planting more climate-resilient crops or crop varieties (Jain et al. 2015). Climate-resilient crops were defined by Dhankher and Foyer (2018) as both crops and crop varieties with enhanced tolerance to biotic and abiotic stresses, which have the ability to maintain or increase crop yields under stressful conditions such as drought, flooding, heat, chilling, freezing, and salinity. Examples of climate-resilient crops are pearl millet, sorghum, and quinoa (Umesh et al. 2019).

Reduce Emissions of Greenhouse Gases

Several practices could be used to reduce the emission of greenhouse gases. These practices may include shifting to conservation tillage which reduced CO_2 emission from the soil. Reducing nitrogen applications without reducing yields include plant tissue testing, soil testing, precision application, use of slow-release fertilizers or nitrification inhibitors, and changes in application timing to better match plant uptake of nutrients (Smith et al. 2008). Intercropping with legumes can reduce the application of nitrogen fertilizer (Zohry et al. 2020), thus it could reduce nitrification.

Producing renewable energy from farm waste and animals manure, namely biogas is also helpful in reducing the emission of CO_2 through burning. In addition, carbon sequestration can be done by practices that increase crop residues left on the field and the volume of organic matter stored in the soil, including planting cover crops, or using additional organic material such as manure (Johnson et al. 2007). Intercropping with legumes can play a role in carbon sequestration in the soil. Abbady et al. (2016) indicated that intercropping pea with sunflower increased carbon sequestration in the soil, compared to the soil cultivated with sole sunflower. Furthermore, planting short-season rice cultivars reduced the amount of methane emitted (Hasan 2013).

IPCC (2014) illustrated the concept of climate-resilient transformation pathways, adapted to the specific case of agriculture, where it clarified that agriculture faces a set of biophysical and socioeconomic stressors, including climate change. Climate-smart pathways result in higher resilience and lower risks to food security, whereas the business-as-usual approach leads to higher risks of changes involving major shifts in agricultural production.

This book tackled the concept of climate-smart agriculture and its three pillars, namely sustainably increases agricultural production, build resilience to climate, and reduce emissions of greenhouse gases. It also contained the existed practices in Egypt that achieve the pillars of climate-smart agriculture. The book also includes some insights on how to improve food production under irrigated conditions, and how to enhance food security under the condition of water scarcity, climate variability, and climate change. Furthermore, the book explained several notions used within the concept of climate-smart agriculture, namely water-smart, soil-smart and crop-smart practices, climate variability, resilient crops, and climate change.

The question that now arises is whether climate-smart practices offset the negative effects of climate variability and change. We tried to answer this question in our book.

Samiha Ouda
Abd El-Hafeez Zohry

References

Abbady K, El-Maaz E, Ahmed H, Zohry AA (2016) Carbon sequestration as a function of intercropping management practice and different nitrogenous fertilizer types. J Soil Sci Agri Engg 7(8):565–586

Asseng S, Ewert F, Rosenzweig C, Jones JW, Hatfield JL (2013) Uncertainty in simulating wheat yields under climate change. Nat Clim Change 3:827–832

Ben-Asher J, Garcia AGY, Hoogenboom G (2008) Effect of high temperature on photosynthesis and transpirationof sweet corn (Zeamays L. var. rugosa). Photosynthetica, 46:595–603

Chaves MM, Zarrouk O, Francisco R, Costa JM, Santos T, Regalado AP, Rodrigues ML, Lopes CM, Costa J, Lopes C (2010) Grapevine under deficit irrigation: Hints from physiological and molecular data. Ann Bot 105:661–676

Dhankher OP, Foyer CH (2018) Climate resilient crops for improving global food security and safety. Plant Cell Environ 41:877–884

Eyshi Rezaei E, Siebert S, Hüging H, Ewert F (2018) Climate change effect on wheat phenology depends on cultivar change. Sci Rep 2018; 8: 4891. 10.1038/s41598-018-23101-2

Fallot A (2016) Témoignagesur la conférence « Climate-smart agriculture 2015 » (Montpellier, 16–18 mars 2015). Nat Sci Sociétés 24:151–153. DOI: 10.1051/nss/2016013

FAO (2010). Climate-smart agriculture: Policies, practices and financing for food security, adaptation and mitigation. Rome: Food and Agriculture Organization of the United Nations

FAO (2013) Climate-smart agriculture sourcebook. Rome: Food and Agriculture Organization of the United Nations (FAO)

FAO (2014) FAO success stories on climate-smart agriculture. Rome: Food and Agriculture Organization of the United Nations (FAO)

Gitz V, Meybeck A (2012) Risks, vulnerabilities and resilience in a context of climate change. In: Building Resilience for Adaptation to Climate Change in the Agriculture Sector; FAO: Rome, Italy, (23) p 19

Hasan E (2013) Proposing mitigation strategies for reducing the impact of rice cultivation on climate change in Egypt. Water Sci (27): 69–77

Hegerl GC, Zwiers FW, Braconnot P, Gillett NP, Luo Y, Orsini JAM, Nicholls N, Hoogenboom G (2000) Contribution of agrometeorology to the simulation of crop production and its applications. Agri For Meteorol 103, 137–157

IPCC (2014) Climate change 2014: Impacts, adaptation, and vulnerability. Part A: Global and Sectoral Aspects. Summary for policymakers (pp. 1–32). Contribution of Working Group II to the Fifth Assessment Report of the Intergovernmental Panel on Climate Change (IPCC). Cambridge University Press, Cambridge and New York, NY

IPCC (2007) Contribution of Working Group II to the Fourth Assessment Report of the Intergovernmental Panel on Climate variability. Cambridge University Press, Cambridge, pp 60

IPCC (2011) Climate change: Impacts, adaptation, and vulnerability. Contribution of Working Group II to the Third Assessment Report of the Intergovernmental Panel on Climate Change. JJ McCarthy, OF Canziani, NA Leary, DJ Dokken, KS White (Eds). Cambridge University Press, Cambridge

IPCC (2012) Managing the Risks of Extreme Events and Disasters to Advance Climate Change Adaptation. A Special Report of Working Groups I and II of the Intergovernmental Panel on Climate Change [Field CB, Barros V, Stocker TF, Qin D, Dokken DJ, Ebi KL, Mastrandrea MD, Mach KJ, Plattner G-K, Allen SK, Tignor M, Midgley PM (eds)].Cambridge University Press, Cambridge, UK, and New York, NY, USA, 582 pp

Jain M, Naeem S, Orlove B, Modi V, DeFries RS (2015). Understanding the causes and consequences of diferential decision-making in adaptation research: adapting to a delayed monsoon onset in Gujarat, India. Glob Environ Change 31: 98–109

Johnson J, Franzluebbers A, Lachnicht WS, Reicosky D (2007) Agricultural opportunities to mitigate greenhouse gas emissions. Environ Poll 150(1):107–24

Khalil AA, Hassanein MK (2016) Extreme weather events and negative impacts on Egyptian agriculture. Int J Adv Res 4(12):1843–1851

Lipper L, Thornton P, Campbell MB, Baedeker T, Braimoh A, Bwalya M, Caron P, Cattaneo A, Garrity D, Henry K, Hottle R, Jackson L, Jarvis A, Kossam Fred, Mann W, McCarthy N, Meybeck A, Neufeldt H, Remington T, Sen TP, Sessa R, Shula R, Tibu A, Torquebiau FE (2014). Climate-smart agriculture for food security. Nat Clim Change 4:1068–1072. 10.1038/nclimate2437

Lobell D, Hammer, G, McLean, G (2013) The critical role of extreme heat for maize production in the United States. Nat Clim Change, 3: 497–501. https://doi.org/10.1038/nclimate1832

Lobell, DB, Monasterio O (2007) Impact of day versus night temperature on spring wheat yield: a comparison of empirical and CERES model predictions in three locations. J Agron 99:469–477

Lobell DB, KG Cassman, Field CB (2009) Crop yield gaps: their importance, magnitudes, and causes. Ann Rev Environ Res 34, 179–204

Mahmood S, Wahid A, Javed F, Basra SMA (2010) Heat stress effects on forage quality characteristics of maize (Zea mays) cultivars. Int J Agric Biol 12:701–706

Martins MA, Tomasella J, Dias CG (2019) Maize yield under a changing climate in the Brazilian Northeast: impacts and adaptation. Agric Water Manag 216, 339–350

Meuwissen MPM, Feindt PH, Spiegel A, Termeer CJ, Mathijs E, de Mey Y, Finger R, Balmann A, Wauters E, Urquhart J, et al (2019) A framework to assess the resilience of farming systems. Agric Syst 176

Motha RP (2011) The Impact of Extreme Weather Events on Agriculture in the United States. *In*: Challenges and Opportunities in Agrometeorology, DOI 10.1007/978-3-642-19360-6_30

Ofori-Sarpong E (2011) Impacts of climate variability on agriculture and farmers' coping strategies in the Upper East Region of Ghana. Department of Geography and Resource Development; University of Ghana, Legon. West Afr J Appl Ecol 2, 21

Ouda S, Hefny Y, Abdel-Wahab S, Abdel-Wahab T (2018). Intercropping patterns of sunflower with peanut under different irrigation regimes and potassium fertilizer levels. Egypt J Agron The 5th International Conference for Crops Science. SI: 85–104

Ouda S, Zohry A, Noreldin T (2018) Crop Rotation Maintains Soil Sustainability. In: Crop Rotation: An approach to Secure Future Food. Springer Publishing House. ISBN 978-3-030-05350-5

Paymard P, Bannayan M, Haghighi RS (2018) Analysis of the climate change effect on wheat production systems and investigate the potential of management strategies

Powell JP, Reinhard S (2016) Measuring the effects of extreme weather events on yields. Weather Clim Extremes 12: 69–79

Rosenzweig C, Iglesias A, Yang X et al (2001) Climate Change and Extreme Weather Events; Implications for Food Production, Plant Diseases, and Pests. Global Change & Human Health, 2:90–104. https://doi.org/10.1023/A:1015086831467

Schlenker W, Roberts JM (2009) Nonlinear temperature effects indicate severe damages to U.S. crop yields under climate change. PNAS, 106 :(37) 15594–15598; https://doi.org/10.1073/pnas.0906865106

Saleh SM, Heggi Mam, Abdrabbo MA, Farag AA (2017) Heat waves investigation during last decades in some climatic regions in Egypt. Egypt J Agric Res, 95(2):863–889

Shah N, Paulsen G (2003) Interaction of drought and high temperature on photosynthesis and grain-filling of wheat. Plant Soil 257:219–226

Smith P, Martino D, Cai Z, Gwary D, Janzen H, Kumar P, McCarl B, Ogle S, O'Mara F, Rice C, Scholes B, Sirotenko O, Howden M, McAllister T, Pan G, Romanenkov V, Schneider U, Towprayoon S, Wattenbach M, Smith J (2008). Greenhouse gas mitigation in agriculture, Philosophical Transactions of the Royal Society. February B(363):789–813

Subedi R, Bhatta DL, Udas E, Agrawal KN, Joshi DK, Panday D (2019) Climate-smart practices for improvement of crop yields in mid-hills of Nepal, Cogent Food & Agriculture 5:1, 1631026, DOI: 10.1080/23311932.2019.1631026

Sweet SKD, Wolfe WA, De Gaetano R, Bennerc (2017) Anatomy of the 2016 drought in the Northeastern United States: Implications for agriculture and water resources in humid climates. Agric For Meteo 247:571–581

Taha A (2012) Effect of climate change on maize and wheat grown under fertigation treatments in newly reclaimed soil. PhD Thesis, Tanta University, Egypt

Umesh MR, Angadi S, Gowda P, Ghimire R, Begna S (2019) Climate-Resilient Minor Crops for Food Security. In: Hasanuzzaman M. (eds) Agronomic Crops. Springer, Singapore. https://doi.org/10.1007/978-981-32-9151-5_2

van der Velde M, Tubiello FN, Vrieling A et al (2012) Impacts of extreme weather on wheat and maize in France: evaluating regional crop simulations against observed data. Clim Change 113:751–765. https://doi.org/10.1007/s10584-011-0368-2

World Meteorological Organization (2012) Standardized precipitation index user guide. Geneva: World Meteorological Organization

Zhang XY, Wang SF, Sun HY, Chen SY, Shao LW, Liu XW (2013) Contribution of cultivar, fertilizer and weather to yield variation of winter wheat over three decades: a case study in the North China Plain. Eur J Agron 50, 52–59

Zohry A, Ouda S (2019) Intercropping systems for sugar beet to improve its land and water productivity. J Soils Crops 29(2):218–226

Zohry AA, Ouda S, Hamd-Alla W, Shalaby E (2017) Evaluation of different crop sequences for wheat and maize in sandy soil. Acta Agriculturae Slovenica 109(2)383–392

Zohry A, Ouda S, Abdel-Wahab T (2020) Sustainable intensive cropping to reduce irrigation-induced erosion: Intercropping systems under surface irrigation practice. Moroccan J Agric Sci 1(2):63–71

Contents

Part I Sustainably Increases Agricultural Production

1 Water-Smart Practices to Manage Water Scarcity 3
Samiha Ouda and Abd El-Hafeez Zohry
1 Introduction 3
2 Water Scarcity Indicators 4
2.1 Falkenmark Indicator 4
2.2 Criticality Ratio of Water 4
2.3 The Classification of the International Water Management Institute 5
3 Water Stress Effects on Crops 5
3.1 Effect of Water Stress on Wheat 5
3.2 Effect of Water Stress on Maize 6
3.3 Effect of Water Stress on Rice 6
3.4 Effect of Water Stress on Sugarcane 7
4 Water Resources in Egypt 7
4.1 The Nile River 8
4.2 Groundwater 9
4.3 Rain Fall 9
4.4 Recycled Agricultural Drainage Water 9
4.5 Total Annual Water Resources of Egypt 10
5 Water Scarcity in Egypt 10
6 Agriculture in Egypt 11
6.1 Irrigated Agriculture Using the Nile River 11
6.2 Irrigated Agriculture by Groundwater 12
6.3 Rain Fed Agriculture 13
7 Causes of Water Scarcity in Egypt 13
7.1 Rapid Growth of Population 13
7.2 Mismanagement of Water Resources and Low Irrigation Efficiency 13

7.3 Agricultural Expansion in the New Lands ... 14
7.4 Pollution ... 14
7.5 Variability in Rain Fall in the Northern Cost of Egypt ... 15
8 Water-Smart Practices to Relief Water Scarcity ... 16
8.1 Irrigation Water Management on Mesqa Level ... 16
8.2 Changing Cultivation/Irrigation Systems ... 17
8.3 Reduction in the Area of High Water Consuming Crops ... 17
8.4 Train Farmers on Irrigation Scheduling ... 17
8.5 Application of Deficit Irrigation ... 18
8.6 Application of Supplementary Irrigation in Rain Fed Area ... 18
9 Water-Smart Practices to Increase Food Production ... 19
9.1 Lands Consolidation ... 19
9.2 Precise Land Leveling ... 19
9.3 Raised Beds Cultivation ... 19
9.4 Implementing Intercropping Systems ... 19
9.5 Improved Water Management Practices of Rain Fed Crops ... 20
9.6 Simulation of the Effect of Application Supplementary Irrigation in Rain Fed Area ... 20
10 Conclusion ... 21
References ... 21

2 Soil-Smart Practices: Integrated Soil Fertility Management ... 27
Abd El-Hafeez Zohry and Samiha Ouda
1 Introduction ... 27
2 Fertilizer Application ... 28
2.1 Chemical Fertilizer ... 29
2.2 Nonchemical Fertilizer ... 30
2.3 Biofertilizer ... 31
3 Integration of Different Fertilizer Sources ... 31
4 Integration Fertilizer Management for Edible Oil Crops: Evidence from Egypt ... 31
5 Case Study of East El-Owinate Region ... 32
6 Effect of Application of Chemical Fertilizer ... 34
7 Effect of Application of Nonchemical Fertilizer ... 34
8 Effect of Application of Integrated Fertilizer ... 35
8.1 Application of 50% Chemical Fertilizer and 50% Biofertilizer ... 36
8.2 Application of 50% Cow Manure and 50% Biofertilizer ... 37
8.3 Application of 50% Compost and 50% Biofertilizer ... 37
9 Conclusion ... 38
References ... 39

3 Fish Farms Effluents for Irrigation and Fertilizer: Field and Modeling Studies 43
Abd El-Hafeez Zohry and Samiha Ouda
1 Introduction 43
2 Studies on Using Fish Farms Effluents for Irrigation in Egypt 45
3 Benefits of Irrigation with Fish Farms Effluents and Implementing Intercropping Systems 45
4 Integration of Water-Smart and Soil-Smart Practices 46
5 Applied Irrigation Water to the Studied Interplanting Systems 46
5.1 Applied Irrigation Water to the Field Crops 48
5.2 Applied Irrigation Water to Interplanted Orange 48
5.3 Applied Irrigation Water to the Studied Crop Sequences 49
6 The Yield of the Studied Field Crops 50
6.1 Winter Crops 50
6.2 Summer Crops 51
6.3 Orange Yield 52
7 Water and Land Equivalent Ratios of the Studied Crop Sequences 53
7.1 Water Equivalent Ratio (WER) 53
7.2 Land Equivalent Ratio (LER) 54
8 Estimation of Potential Irrigated Lands by Fish Farms Effluents 56
9 The Suggested Cropping Systems 58
9.1 Crop Rotations 58
10 Calculation of Water Requirements of the Suggested Crop Rotations 61
10.1 Interplanting Legume Crops Between Orchard Trees 61
11 Conclusion 63
References 63

4 Integration Between Crop-Smart, Water-Smart and Soil-Smart Practices 67
Abd El-Hafeez Zohry and Samiha Ouda
1 Introduction 67
2 Intercropping Systems 68
2.1 Intercropping Attains Yield Advantage 69
3 Land Equivalent Ratio 69
3.1 Higher Water Productivity Under Intercropping Systems 70
4 Water Equivalent Ratio (WER) 71
4.1 Intercropping Realizes Soil Conservation 71
4.2 Intercropping Control Crop Predators' Infestation 73
4.3 Intercropping Modifies the Microclimate of the Associated Crops 73
5 The Sequence of the Cultivated Crops 74
6 Crop Rotation 75

7 Crop Rotations in Egypt ... 76
7.1 Cotton Rotations ... 76
7.2 Intensive-Cropping Rotations ... 77
8 Calculation of the Applied Irrigation Water and Water Use Efficiency of These Three Rotations ... 80
8.1 Cereal Crops Rotations ... 81
8.2 Legume Crops Rotation ... 82
8.3 Crop Rotations to Defeat Diseases ... 82
8.4 Crop Rotations to Conserve Irrigation Water ... 83
9 Conclusions ... 84
References ... 84

Part II Build Resilience to Climate

5 Climate Extremes and Crops ... 93
Samiha Ouda and Abd El-Hafeez Zohry
1 Introduction ... 93
2 Climate Variability Leads to Extreme Weather Events ... 95
3 Effects of Climate Variability on Crops ... 95
3.1 Effects of Heat Waves ... 95
3.2 Effect of Precipitation Variability ... 96
3.3 Interaction Between High Temperature and Precipitation Variability ... 96
4 Climate Variability and Agriculture in Egypt ... 97
4.1 Climate Variability Effects on Fresh Water Resource ... 97
4.2 Climate Variability and Rain Fall in North Coast of Egypt ... 97
5 Heat Waves and Crops Production in Egypt ... 98
5.1 Wheat Crop ... 98
5.2 Rice Crop ... 99
5.3 Maize Crop ... 99
6 Measures to Limit Climate Variability Stress in Egypt ... 100
6.1 Implementing Resilient Farming System ... 100
6.2 Increasing Crops Diversification ... 102
6.3 Increasing Resilience of Crops/Varieties ... 105
6.4 Early Warning System for Extreme Weather Events ... 107
7 Conclusions ... 107
References ... 107

6 Climate-Resilient Crops ... 115
Abd El-Hafeez Zohry and Samiha Ouda
1 Introduction ... 115
2 Advantage of Using Climate-Resilient Crops ... 117
2.1 The Advantage of Using Quinoa as a Food Crop ... 117
3 Nutritional Advantage of Quinoa ... 117

4 Traits Contributing to Climate-Resiliency of Quinoa 118
4.1 Tolerance to Heat Stress 118
4.2 Tolerance to Salinity Stress 119
4.3 Tolerance to Water Stress 120
5 Quinoa Cultivation in Egypt 121
6 The Advantage of Using Millets as a Food Crop 122
7 Nutritional Advantages of Pearl Millet 123
8 Traits Contributing to Climate-Resiliency of Millets 124
8.1 Tolerance to Heat Stress 124
8.2 Tolerance to Salinity Stress 124
8.3 Tolerance to Water Stress 124
9 Pearl Millet Cultivation in Egypt 125
10 The Advantage of Using Sorghum as a Food Crop 126
11 Nutritional Advantages of Sorghum 126
12 Traits Contributing to Climate-Resiliency of Sorghum 127
12.1 Tolerance to Heat Stress 127
12.2 Tolerance to Salinity Stress 127
12.3 Tolerance to Water Stress 128
13 Sorghum Cultivation in Egypt 128
14 Conclusion 129
References 129

7 Assessment of Climate Variability on Wheat Productivity in Egypt 137
Samiha Ouda
1 Introduction 138
2 The Egyptian Wheat Cultivars 139
3 Effect of Weather Elements on Wheat Growth and Yield 140
3.1 Temperature 140
3.2 Solar Radiation 141
3.3 Relative Humidity 142
4 Methodology of the Analysis 142
4.1 Collected Data 142
5 Statistical Analysis 144
6 Coefficient of Determination (R^2) 144
7 Root Mean Square Error (RMSE) 144
8 Willmott Index of Agreement (*d-stat*) 145
9 Results and Discussion 145
9.1 Assessment of the Spatial-Variability of Wheat Zone-Specific Productivity and Weather 145
9.2 Assessment of Temporal-Variability of National Wheat Productivity and Climate 152
10 Conclusions 160
References 162

Part III Reduction of Greenhouse Gases

8 Practices Contribute in Reducing the Emission of Greenhouse Gases 167
Samiha Ouda and Abd El-Hafeez Zohry
1 Introduction 167
2 Implication of Climate Change on Agriculture 169
2.1 Effects on Water Resources 169
2.2 Effects on Crops Production 169
2.3 Effects on Agricultural Soils 171
2.4 Effects on Water Consumption by Crops 171
3 Implication of Climate Change in Egypt 172
3.1 Climate Change and Water Resources 172
3.2 Climate Change and Water Needs for the Cropped Area 173
3.3 Climate Change and Crops Production 174
4 Source of Emission of Greenhouse Gases 174
4.1 Emission from the Soil 174
4.2 Emission from Rice Fields 175
4.3 Burning of Crops Residues 176
5 Climate-Smart Practices to Reduce Emission of Greenhouse Gases 176
5.1 Carbon Sequestration in the Soil 176
5.2 Reduction of Greenhouse Gases Emissions from Rice Fields 177
5.3 Biogas Production 177
6 Conclusion 178
References 178

Part I
Sustainably Increases Agricultural Production

Chapter 1
Water-Smart Practices to Manage Water Scarcity

Samiha Ouda and Abd El-Hafeez Zohry

Abstract Increasing world population resulted in increasing the demand on water for food production, which resulted in water shortage/scarcity. Water scarcity is the point at which the aggregate impact of all users impinges on the supply or quality of water under prevailing institutional arrangements to the extent that the demand by all sectors, including the environment, cannot be fully satisfied. In this chapter, we overviewed water scarcity indicators, and effect of water scarcity on the production of several crops. Furthermore, water resources and agriculture in Egypt (irrigated and rain fed) was overviewed. The causes of water scarcity in Egypt and water-smart practices that contribute to relief water scarcity, as well as increase food availability were discussed.

Keywords Water scarcity indicators · Effect of water stress on crops · Water resources in Egypt · Causes of water scarcity in Egypt

1 Introduction

There is a challenge in arid and semiarid regions owing to fresh water shortage, wide climate variability, and population growth (Forouzani and Karami 2011). The United Nations (2006) defined water scarcity as "the point at which the aggregate impact of all users impinges on the supply or quality of water under prevailing institutional arrangements to the extent that the demand by all sectors, including the environment, cannot be fully satisfied. It is a relative concept and can occur at any level of supply or demand". Whereas, FAO (2012) described water scarcity with a simple definition as: an excess of water demand over available supply, which exists as a consequence of a high rate of aggregate demand from all water-using sectors compared with available

S. Ouda (✉)
Water Requirements and Field Irrigation Research Department, Soils, Water and Environment Research Institute; Agricultural Research Center, Giza, Egypt

A. E.-H. Zohry
Crops Intesefication Research Department, Field Crops Research Institute; Agricultural Research Center, Giza, Egypt

S. Ouda and A. E.-H. Zohry, *Climate-Smart Agriculture*,
https://doi.org/10.1007/978-3-030-93111-7_1

supply, under the prevailing institutional arrangements and infrastructural conditions. Both reports characterized three main dimensions affecting water scarcity, namely a physical lack of water availability to satisfy demand; the failure of infrastructure development that controls storage, distribution and access; and low institutional capacity to provide the necessary water services.

Increasing world population in an exponential rate between 1961 and 2000 resulted in increasing the demand on water for food production (Keating et al. 2014). The largest fresh water user on the planet is agriculture, where it consumes more than two thirds of total withdrawals (Gan et al. 2013). Consequently, fresh water resources available to agriculture will need to be used with wisdom to satisfy the developmental needs of other sectors (Chai et al. 2016). In arid and semi-arid regions, irrigated agriculture is crucial to attain stable agricultural production, it aids crop intensification and diversification (Vlek et al. 2008), it facilitate the application of other intensification inputs, namely fertilizers and chemicals via irrigation water (Kögler and Söffker 2017).

In this chapter, we overviewed water scarcity indicators, effect of water scarcity on the yield of several crops, water resources and agriculture in Egypt (irrigated and rain fed), and the causes of water scarcity. Additionally, we discussed suggested water-smart practices to relief water scarcity and to increase food availability.

2 Water Scarcity Indicators

Several water scarcity indicators are used in the literature to measure the condition of water scarcity as followed.

2.1 Falkenmark Indicator

Falkenmark indicator is the most commonly used in the literature and it is also called water stress index (Falkenmark et al. 1989). It defines water scarcity as the amount of renewable freshwater that is available for each person each year in a certain country. It indicated that a country is consider to experience water stress if the amount of renewable water is below 1,700 m^3/person/year. If it is below 1,000 m^3/person/year, the country is experiencing water scarcity; and if it is below 500 m^3/person/year, the country is experiencing absolute water scarcity.

2.2 Criticality Ratio of Water

Another indicator used in defining and measuring water scarcity developed by Raskin et al. (1997) is the "criticality ratio". This approach measures water scarcity as the proportion of total annual water withdrawals relative to total available water resources in a country. Using this approach, a country is said to be water scarce if annual withdrawals are between 20–40% of annual supply, and severely water scarce if they exceed 40%.

2.3 *The Classification of the International Water Management Institute*

A third measure of water scarcity was developed by the International Water Management Institute (IWMI). In this approach, the IWMI classifies countries that are predicted to be unable to meet their future water demand without investment in water infrastructure and efficiency as economically water scarce; and countries predicted to be unable to meet their future demand, even with such investment, as physically water scarce (Molden 2007).

3 Water Stress Effects on Crops

Water stress can affect all morphological and physiological processes of the crop. The general response is reflected in a reduction in plant size, leaf area and harvestable yield (Sadras et al. 2016). Water stress induces a decrease in leaf water potential and in stomatal opening, which reduced availability of CO_2 for photosynthesis (Osakabe et al. 2013).

3.1 *Effect of Water Stress on Wheat*

Water stress differently affects the growth of wheat at different growth stages by restricting photosynthesis in winter wheat (Li et al. 2015). Liu et al. (2016) indicated that a progressive inhibition of photosynthetic activity was observed as water stress became more severe in field-grown bread wheat, which led to lower grain yields even at moderate water stress. Low soil water potential causes a progressive reduction in leaf relative water content, water potential (Habash et al. 2014), reduces plant height, biomass and yield of wheat (Li et al. 2015). Mirzaei et al. (2011) reported that drought stress at all growth stages induced reducing grain yield and yield components.

During vegetative growth and under water stress, phyllochron decreases in wheat and leaves become smaller, which could reduce leaf area index (Gupta et al. 2001) and number of reproductive tillers, in addition to limit their contribution to grain yield (Dencic et al. 2000). Furthermore, sensitivity for water stress is the highest in stem elongation stage than in other growth stages (Gupta et al. 2001). Whereas,

Kirigwi et al. (2004) indicated that the most sensitive growth stage to lack of water is the flag leaf stage, followed by the flowering stage. Water shortage in the flag leaf stage reduces the assimilation rate and accelerates the decomposition of assimilation during senescence (Farkas et al. 2020).

Zareian et al. (2014) reported that water stress during ear emergence and grain filling phases reduced grain yield and its components. Mirzaei et al. (2011) reported that drought stress at stages of stem elongation, flowering and grain filling stages induced 32, 32 and 35% reduction in grain yield, respectively. Bauder (2001) indicated that drought stress during maturity resulted in about 10% decrease in yield.

3.2 *Effect of Water Stress on Maize*

Water stress affects maize vegetative and reproductive growth causing high yield losses. Early season drought reduces plant growth and inhibits plant development (Heiniger 2001). Grzesiak (2001) reported that early stage of seedling growth and establishment is very sensitive to drought. Drought stress during vegetative growth stages especially during V1–V5, reduces growth rate, prolong vegetative growth stage and conversely the duration of reproductive growth stage is reduced (Aslam et al. 2013). A comparison was made between maize yield resulted from low irrigation water application during the vegetative stage and during the whole growing season revealed that yield losses were increased by 10–20% in the first case, compare to the second case (Domínguez et al. 2012).

A short duration of water deficit during tasselling stage in maize resulted in reduction in biomass production and grain yield by 30 and 40%, respectively (Çakir 2004). Water stress increased the days to anthesis, days to maturity, decreased the leaf number, and loss of normal root architecture, which all lead to a reduction in maize grain yield (Sah et al. 2020). Drought occurring between two weeks before and two weeks after silking stage can cause significant reductions in kernel set and kernel weight resulting in an average of 20–50% yield loss (Nielsen 2007). Kamara, et al. (2003) reported that water deficit reduced biomass accumulation by 37% at silking, by 34% at grain filling period, and by 21% at maturity. Loss of grain production under drought in maize is connected to shortages of water used for irrigation (Pandit et al. 2017). The loss of yield varies from 30 to 90% depending on the crop stage, as well as the degree and duration of water stress (Pandit et al. 2018).

3.3 *Effect of Water Stress on Rice*

Water shortage is a very important factor in rice production, not only the amount of water deficit but also the time of deficit. When rice plants are subjected to water stress, a varieties of physiological changes are occurred, such as reduction of transpiration rate, chlorophyll content in leaves, photosynthetic rate, stomatal conductance, and relative water content resulting in decrease of plant growth (Jahan et al. 2013). During the early vegetative growth and flowering of rice, exposure to water stress drastically inhibits floret initiation, plant height, tiller number, leaf area, grain filling and ultimately causing low grain production (Yang et al. 2019). Water stress tends to delay flowering, and a larger delay in flowering was associated with a higher reduction in grain yield, harvest index and percentages of fertile panicles and filled grains (Pantuwan et al. 2000). Drought for two weeks from 48 to 62 days after transplanting using withholding irrigation, significantly reduced rice grain yield (Ravindra et al. 2002). Water deficit from booting to grain filling stage caused greatest rice yield reduction by 77%, compared to water stress during all the growth stages (vegetative, panicle initiation and boot to grain fill), which also reduced yield components (Harbir and Ingram 2000). Rice growth period from booting to grain fill (reproductive stage) was the most sensitive to water deficit (Harbir et al. 2002).

3.4 Effect of Water Stress on Sugarcane

The susceptibility of sugarcane to water stress is greater in the tillering and stem elongation phases (Machado et al. 2009) with both stem and leaf growth are the most affected than other organs (Lakshmanan and Robinson 2014). This is mainly because 70–80% of sugarcane yield is produced during this phase (Singh and Rao 1987). Ethan et al. (2016) indicated that water deficit during this stage significantly decreased sugarcane yield. The most common water stress responses in sugarcane are leaf rolling, stomatal closure, inhibition of stalk and leaf growth, leaf senescence and reduced leaf area (Inman-Bamber et al. 2012). Moreover, under water stress, both cell division and cell elongation are interrupted (Machado et al. 2009) and stem and leaf elongation are the most seriously affected growth processes (Inman-Bamber et al. 2008). Root development is also influenced by water deficit, but with relatively less effect than the above-ground biomass (Smit and Singels 2006).

4 Water Resources in Egypt

Egypt is located on the Northeastern corner of Africa on the Mediterranean Sea between latitudes 22° and 32° N, and between longitudes 24° and 37° E. The total area of Egypt is 1,001,450 km^2, with a land area of 995,450 km^2. The Egyptian terrain consists of vast desert plateau interrupted by the Nile Delta and Valley, which occupy about 4% of the total area of Egypt. The Nile Delta and Valley divided the desert land of Egypt into the western desert (represents two third of Egypt territory) and the eastern desert. In addition, Sinai Peninsula located in the eastern part of Egypt represents 6% of the Egypt area (Zohry and Ouda 2020). Sinai Peninsula is located in Asia continent, which makes Egypt a transcontinental country (Omran

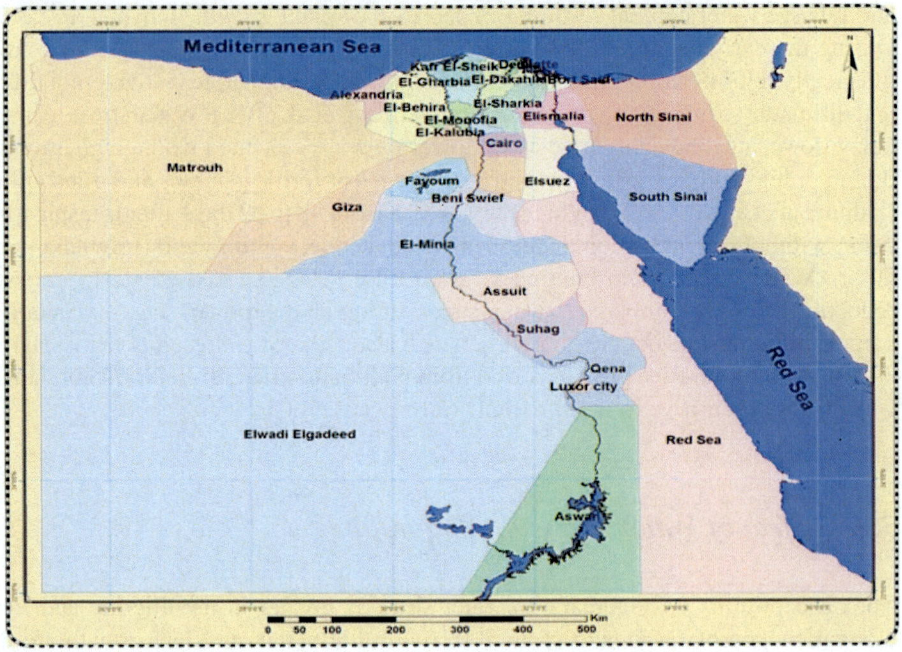

Fig. 1 Map of Egypt

2017). Egypt is bordered by Libya in the west, by the Gaza Strip and the Red Sea in the east, and by Sudan in the south.

The inhabited area of the country is confined to the Nile Delta and Valley. The Nile Delta covers the area from Cairo to the shoreline of the Mediterranean Sea, between the cities of Damietta in the east and Rashid in the west. A narrow strip from south of Cairo to Aswan represents the Nile Valley (Morsy et al. 2017) (Fig. 1).

Water resources in Egypt are very limited, where more than 95% of it is received from outside of its international borders. It is consists of Egypt water share from the Nile River, ground water, effective rainfalls on the northern strip of the Nile Delta, and recycled agricultural drainage water.

4.1 The Nile River

The Nile River is the main water resource in Egypt, where Egypt lies at the end of the Nile's route toward the sea. The annual share of Nile River water was determined in an international agreement by 55.5 billion cubic meters. The High Aswan Dam is the major regulatory facility on the river. It has a capacity of 162 billion cubic meters (including 90 billion cubic meters of live storage). The high Aswan Dam is completely eliminating the flood risks and allowing year-round cultivation. The annual share of the Nile is stored in Lake Nasser. The Nile water is driven from the dam downstream by intensive network of canals through several types of control structure (Ministry of Water Resources and Irrigation 2005).

4.2 Groundwater

Groundwater is the second most important resource of water after the Nile River. Ground water is the portion of the water beneath the surface of the earth that can be collected with the wells, tunnels, drainage galleries, or that flows naturally to the earth's surface via seeps or springs (Ashour et al. 2010). There are different groundwater aquifers with variable importance for exploitation in the Nile river region. The first category of the groundwater is shallow local aquifers recharged by rainfall comprises groundwater in the Nile Valley and Delta system. The other category is deep not replenish aquifers, which are non-renewable type and located in the Western Desert-Nubian Sandstone Aquifer (Abdel-Shafy and Kamel 2016). The current total abstraction of the groundwater is only 2.0 billion cubic meters per year as a result of great depth, up to 1500 m in some area (Ministry of Water Resources and Irrigation 2013).

4.3 Rain Fall

The third important water resource in Egypt is rain, which falls in the winter only. The amount of rain in the Egyptian north coast varies ranging from 130 to 150 mm in the northwestern coast and from 80 mm (west of Al-Arish) to 280 mm (at Rafah) in the northeastern coast. This rate decreases after 20 km south of the Mediterranean in both areas. The effective rainfalls on the northern strip of the Nile Delta are calculated by 1.3 billion cubic meters per year (http://www.emwis-eg.org). The rain fall in Egypt cannot be considered as a reliable source for water.

4.4 Recycled Agricultural Drainage Water

The reuse of agricultural drainage water in irrigation has been adapted as an official policy since the late seventies by mixing it with Nile water (fresh water) in the main and branch canals.

4.5 Total Annual Water Resources of Egypt

The total annual water resources of Egypt is composed of 55.5 billion cubic meters obtains from the Nile, 2.0 billion cubic meters from ground water, 1.3 billion cubic meters from rain, which add to 58.8 billion cubic meters. The major amount of water consumption goes to agriculture, namely about 62.3 billion cubic meters. Drinking water uses 8.7 billion cubic meters and industry uses about 7.5 billion cubic meters. These three amounts add up with total of 78.5 billion cubic meters. Thus, there is a gap estimated by about 20 billion cubic meters between available water resources and the demand for water. This gap is overcome by recycling agricultural drainage water to be used in irrigation (Ministry of Water Resources and Irrigation 2014). It is only used to fill the gap between water supply and demand and not considered as independent resource.

5 Water Scarcity in Egypt

Egypt is currently using more water than its internal renewable resources supply and is expected to use even more water in the near future. According to the Egyptian Statistics Agency (CAPMAS), the country's per capita water resources have fallen from 2,526 m^3/year in 1947 to less than 700 m^3/year in 2013, which is well below the 1,000 m^3/year threshold deemed necessary by the United Nations to provide enough water for drinking, agriculture and nutrition (United Nations 2006). This trend is expected to continue, leading to a possible figure of less than 350 m^3/year by 2050 (UNEP 2014). Similarly, aggregate water consumption is expected to increase by more than 20% in the next few years (Ministry of Water Resources and Irrigation 2014), while upstream development projects on the Nile pose as a risk to reduce the amount of water flowing down to Egypt.

The official water policy is to satisfy the needs of drinking water and industry first and what is left is assigned to agriculture. Putting agricultural water needs last present physical lack of water availability. Although Egypt has effective infrastructure development that controls storage, namely the High Aswan Dam, the distribution network is not as effective to make access to irrigation water easily. Furthermore, low institutional capacity to provide the necessary water services is also exist (Ashour et al. 2010).

6 Agriculture in Egypt

6.1 Irrigated Agriculture Using the Nile River

Irrigated agriculture is the main source for food production in Egypt. It uses 85% of all the Egyptian water resources. The Nile River is the main source of irrigation in the old lands of Egypt. The Nile's water is diverted from Aswan High Dam through eight river barrages to a 31,000 km network of canals. Irrigation canals are classified into principal canals (Rayah), main canals (first level canals), branch canals (second level canals), distribution canals (Mesqas, or third level canals) and irrigation ditches (Merwas) (Allam 2004). Flow in the principal and main canals is continuous, whereas in the branch and distribution canals the run is on a rotation basis. Furthermore, the government is supporting a process of converting some of the lower-level canals gradually to continuous flow (Ministry of Water Resources and Irrigation 2006). Figure 2 shows a sketch of the system of barrages and main canals along the Nile.

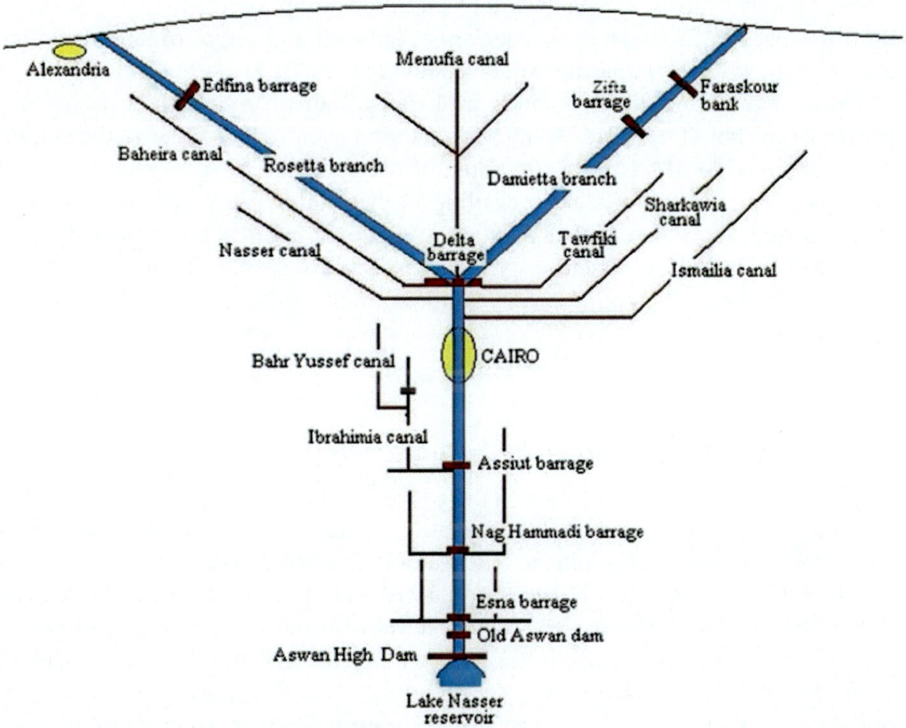

Fig. 2 Sketch of the system of barrages and main canals along the Nile. *Source* R. J. Oosterbaan at nl.wikipedia—transferred from nl.wikipedia (Original text: eigen werk), Public Domain, https://commons.wikimedia.org/w/index.php?curid=7321839

Almost all irrigated land in Egypt uses machinery to lift water from waterways to the level of the fields, with few exceptions include Fayoum governorate. A general system of "water duties" in the Nile Delta and Valley are 5 days on and 10 days off in the winter, with a winter closure period in January. While, in the summer, the water duties are 5 days on and 5 days off.

According to CAPMAS (2018), the area of irrigated lands in Egypt is composed of 3.78 million hectare in the old lands, which are located in the Nile Valley and Delta Regions. It is characterized by having alluvial soils (clay to loamy). The old lands is under surface irrigation, with few areas converted to drip systems. Irrigation application efficiency of surface irrigation is between 50 and 60%, depending on the cultivated crop. The total water requirements for the cultivated crops in the old lands were estimated by Ouda and Zohry (2018a) to be 49.5 BCM.

In the old lands, farmers are used to cultivate their crops in basins, where they broadcast seeds, or they plant seeds on furrows. Under furrow irrigation, water runoff could cause soil erosion and loss in fertility (Sojka et al. 2007). In addition, Zohry et al. (2020) indicated that lower productivity of wheat and sunflower was found under furrow irrigation. Either cultivation in basins or on furrows results in application of a large amount of irrigation water and result in low water use efficiency. In addition, fertilizer leaching is also a consequence of application of large irrigation amounts, causing groundwater pollution (Abouelenein et al. 2010).

On the other hand, the new lands are located mainly on both the east and west fringes of the Nile Delta and Valley and scattered over various areas in the country. According to CAPMAS (2018), the area of the new lands is 1.53 million hectare. Sandy and calcareous soils are prevailing in these areas. Some of these areas are irrigated from the Nile canals expended in this area and the rest is irrigated with groundwater. Irrigation systems (sprinkler or drip) are used in these types of soils. Water application efficiency is 75–80% for sprinkler system and 85–90% for drip system.

6.2 *Irrigated Agriculture by Groundwater*

Groundwater is the main source of irrigation water in the cultivated new lands of most of the governorates, where no Nile water is available. In the new lands, the total water requirements for the cultivated crops were estimated by 12.9 BCM (Ouda and Zohry 2018b). The Egyptian government initiated many agricultural developments projects in the Western desert using groundwater for irrigation, namely Tushky in Upper Egypt and East El-Owinat project in southeastern Egypt. Another development project was initiated in 2015 to reclaim 1.5 million acres in 16 different locations of Egypt, where most of the lands are located in the Western desert region with the primary goal of increasing agricultural production (Moghazy and Kaluarachchi 2020). In these areas, the main irrigation system is the pivot system, in addition to sprinkler or drip systems.

6.3 Rain Fed Agriculture

Rain fed agriculture exists in the northern coast of Egypt, where barley is the main cultivated. Wheat is also cultivated, in addition to some fruit trees. Khalifa et al. (2004) indicated that, in the northeastern coast of Egypt, barely was cultivated every year, which resulted in low productivity. Whereas, in the northwestern coast of Egypt, wheat is cultivated and its productivity was low due to its cultivation every year (Attia and Barsoum 2013). Governmental interventions in crop management practices are now implemented to replace the traditional crop production and primitive soil management with more contemporary management to attain better yield of the cultivated crops in these areas.

7 Causes of Water Scarcity in Egypt

Several setbacks exist in Egypt causing water scarcity in the recent years. Some of these factors are: population increase, misuse of water resources, inefficient irrigation techniques, expansion in the new lands, pollution causing environmental degradation and variability in rain fall in the northern cost of Egypt.

7.1 Rapid Growth of Population

According to the website "WorldoMeter", Egypt ranks number 14 in the list of countries by population. Egypt population is equivalent to 1.31% of the total world population and population density is 103 per Km^2. High population increase causes two main problems, namely enlarges water stress by boosting domestic water consumption and increases irrigation water use to meet higher food demands (Dakkak 2016). Egypt has one of the highest population growth rates in the Middle East. The number of people living in the country has doubled from 35.3 million to 69.8 million between 1970 and 2001 (UNEP 2014). United Nations projected an increase in population to 95.6 million by 2026, Egypt passed this figure and reached 100 million in 2019 (CAPMAS 2019). Furthermore, Alkitkat (2017) projected 32% increase in the Egyptian population by the year of 2030, compared to its value in 2014.

7.2 Mismanagement of Water Resources and Low Irrigation Efficiency

Wasteful irrigation practices, and deficient water delivery infrastructures are the factors reducing the amount of available water in Egypt. Only 6% of the total irrigated

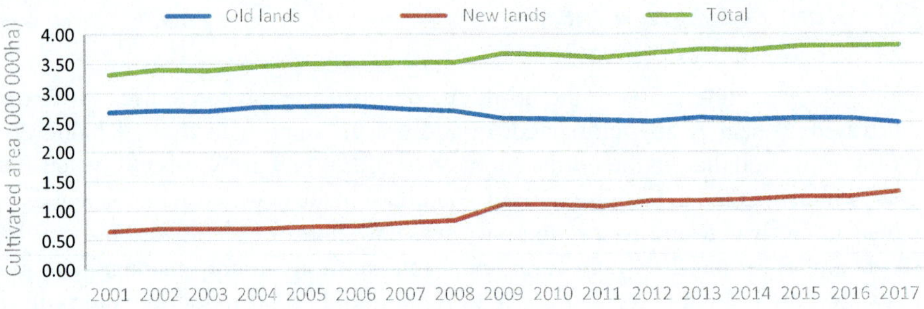

Fig. 3 The cultivated area in Egypt from 2001 to 2017

areas use improved irrigation systems. Furthermore, Ashour et al. (2010) stated that severe shortage of water at the tail reaches, low conveyance efficiency, high operational cost, and excessive water losses to drains occur, which elevate the level of water scarcity situation in Egypt. Additional water is lost due to leaking pipes and drains (El Quosy 2006). Liberalization of cropping pattern on farm level as a result of the ramified and inter-connected nature of the water distribution network also is another cause of mismanagement of water (Molle 2019). Unpredicted timing and quantity of available water results, not only in a mismatch between supply and demand, but also resulted in over-irrigation or sooner irrigation than required by the farmers as a means of storing water in the soil profile to minimize the risk of not irrigating at all (El Quosy 2006).

To reduce the damage of these above obstructions, the government of Egypt have implements several projects for irrigation system modernization.

7.3 Agricultural Expansion in the New Lands

An accelerated increase in the expansion in the new lands was observed from 2001 to 2017 with 104% increase (Fig. 3). Meanwhile, area of the old lands were increase until 2009, then it started to decrease. Most of these new areas are irrigated with groundwater and the some are irrigated from the Nile water. This put pressure on water resources and have contributed in water scarcity.

7.4 Pollution

The problem of the river Nile pollution has been regularly underestimated. There are many economic sectors relying on the Nile, namely drinking, agriculture, and municipal use. Due to a lack of water treatment facilities and regulations, agricultural runoffs containing pesticides, are industrial effluents are, in some cases, being dumped in the Nile River (El Bedawy 2014). These are of significant risk not only on human health, but also on animal health and agricultural production (Dakkak 2016). Agricultural production quality and quantity has been affected by using untreated water for irrigation as the bacteria and the metals in the water affect the growth of the plant produce, especially in the Nile Delta where pollution is the highest. El Tahlawi et al. (2008) summarized the most environmental problems in the Nile Delta as pollution of surface water, northern lakes, the shallow groundwater layers and aquifers due to the disposal of primary or non-treated waste water, and solid waste disposal, as well as the over-pumping on the desert fringes caused decline of groundwater level.

Of course the pollution of Nile is a complex problem, but the government of Egypt, recently, has implemented stricter rules to reduce the pollution of the Nile and identified the quality and the type of waste/wastewater dumped into the river. Furthermore, rapid and significant actions has been done to overcome such drawbacks.

7.5 Variability in Rain Fall in the Northern Cost of Egypt

High variability of rain fall are exist in the northern cost of Egypt as it is shows from the available rain fall data in the period from 1997 to 2020 (Fig. 4). The figure showed that highest value of rain fall was found in 2020, namely 276 and 376 mm for northeastern and northwestern costs, respectively. The lowest values were was found in 2019, namely 70 mm, for each of northeastern and northwestern coasts with range between the highest and the lowest values equal to 206 mm.

The level of dispersion around the mean was high as it presented by high CV%, namely 30 and 31% for northeastern and northwestern costs, respectively.

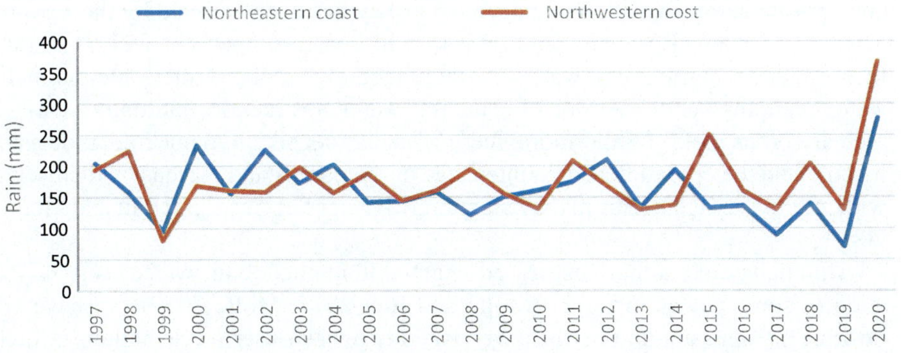

Fig. 4 Annual values of rain fall (mm) from 1997 to 2020 in northeastern and northwestern costs of Egypt

8 Water-Smart Practices to Relief Water Scarcity

Efficient irrigation water management is the solution to relief the stressful conditions of water scarcity. Several procedures can be very effective to reduce water losses and increase availability of water for the cultivated crops to complete their life cycle without water stress. Furthermore, the saved water amounts could allow cultivation of new areas and increase food production.

8.1 Irrigation Water Management on Mesqa Level

Water run through a Mesqa, which is typically serves an area of 20–80 ha (Allam 2004). Improved Mesqas (concrete lined) is the major aspect in improving water management. Other aspects are down-stream control devices, single lifting points, Irrigation Advisory Services, and continuous flow in distributary's canals and their branches. Continuous flow means the existence of water in the irrigation system all the time and avoiding the periodical old system (Ashour et al. 2010).

A project call Farm-level Irrigation Modernization Project was implemented in Egypt to increased access to modern irrigation systems in the Mahmoudia, Manaifa and Meet Yazid canals in Egypt, primarily by phasing out diesel pumps and introducing electric ones, as well as by constructing an electricity grid to power the modernized pumping stations. A total of 197,633 water users (including landholders and tenants) benefited through improved irrigation and drainage services in this project (The World Bank 2020).

In the beginning of the 1990s, the government began to transfer the responsibility for the management of branch canals to water user's associations, in a process called "irrigation management transfer" (El Quosy 2006). Water user's associations are a participatory irrigation management policy has been adopted by the Egyptian Ministry of Water Resources and Irrigation to decrease open-channel conveyance losses, increase agricultural water distribution equity among farmers and rationalize water pumping, with the aim of reducing water abstraction, pumping costs and increase crops yield. Furthermore, these associations play a major role in decision-making and the operation and maintenance of the pumps and mesqas by themselves, with minimal assistance from the Irrigation Advisory Service staff (Batt and Merkley 2009).

Although some authors expressed negative thoughts about water users' association in some places in Egypt (Molle and Rap 2014; Molle 2019), we have very positive thoughts about it because we visited two water user's associations in Luxor governorate, south Egypt in 2020. We observed how the members of the association expressed their happiness for being able to implement the improvement in the mesqa under the supervision of the Ministry of Water Resources and Irrigation. They said that the improvement reduced irrigation time and saved water and fuel.

8.2 Changing Cultivation/Irrigation Systems

Changing cultivation system from basins or furrows to raised beds proved to save on the applied irrigation water. Zohry et al. (2020) indicated that 13–15% of the applied water to wheat was saved under raised beds cultivation, compared to cultivation on furrows. Cultivation on raised beds exerts control on the applied water resulting in less water and nutrient loss through deep percolation, thus reduced total water requirements for crops (Dogan and Kirnak 2010). Ouda and Zohry (2018b) reported that expanding on raised beds cultivation for the entire prevailing cropping pattern in Egypt could save 9.9 billion cubic meters of the applied irrigation water under furrow irrigation.

Furthermore, changing irrigation system from surface (60% application efficiency) to more efficient system, namely sprinkler (75–80% application efficiency) or drip systems (85–90% application efficiency) could result in saving lager amounts of the applied irrigation water. Under these circumstances, technical advices from extension workers should be provided to prevent soil deterioration.

8.3 Reduction in the Area of High Water Consuming Crops

Egypt cultivates two crops that have high water requirements, namely rice in the Nile Delta and sugarcane in Upper Egypt. Reduction in the cultivated area of these two crops can help in saving some of applied irrigation water and reduce water scarcity situation.

8.4 Train Farmers on Irrigation Scheduling

One of the most important factors that contribute in water scarcity situation in Egypt is the wasteful use of irrigation water by the farmers. Simple irrigation scheduling technique for the cultivated crops should be taught to the farmers to help them to apply irrigation on the time that the growing plants need.

Irrigation scheduling is a practice aimed to achieve an optimum water supply for better productivity, with soil water content being maintained close to field capacity (Tariq and Usman 2009). Furthermore, the aim of irrigation scheduling is to maximize yield, enhance irrigation efficiency, and improve crop quality by applying the exact amount of water needed by the crop, or to replenish the soil moisture to the desired level (Ali 2010). Thus, reduce irrigation water losses to the groundwater and diminish water scarcity situation. Taha and Ouda (2016) indicated that irrigation scheduling for wheat and maize contributed in saving 25 and 28% of the applied irrigation water, compared to the amount applied by the farmer.

8.5 *Application of Deficit Irrigation*

Deficit irrigation is one of the most important management strategies. Fereres and Soriano (2007) defined deficit irrigation as an irrigation strategy to maximize yield with a minimum rate of water application. Deficit irrigation increases water use efficiency through increase in application efficiency, consumption efficiency and yield efficiency (Hsiao et al. 2007). Increases in application efficiency occur as a result of lower amount of water applied than full evapotranspiration, thus most or all the water applied remains in the root zone and water lost by run-off and deep percolation decreases (Sepaskhah and Ghahraman 2004). Several strategies of deficit irrigation could be practice, namely sustained deficit irrigation (Fernandes-Silva et al. 2018), regulated deficit irrigation (Fabeiro Cortés et al. 2003) and subsurface drip irrigation (Chai et al. 2016).

Ouda et al. (2020) assessed the potential effect of individual cultivation of four winter crops grown in Egypt (Egyptian clover, sugar beet, onion and tomato) on raised beds, instead of furrow cultivation and application of sustained deficit irrigation on national level. They modeled the increase in the productivity of these four crops under cultivation on raised beds and the decrease in yield under deficit irrigation. They concluded that although losses in the national production of these four crops occurred, it did not affect their self-sufficiency ratio. Furthermore, modeled the use of the saved amount of irrigation water in cultivate new areas with wheat on the national level (its self-sufficiency ratio is 43%). They found that this procedure could increase wheat self-sufficiency ratio to 80%.

8.6 *Application of Supplementary Irrigation in Rain Fed Area*

Application of supplementary irrigation to increase productivity of cultivated crops under rain fed conditions can be implemented if macro-catchment water harvesting techniques were practiced. Hussain (2005) indicated that application of 30 mm as supplementary irrigation during heading stage increased yield of wheat by 30% in the northeastern coast of Egypt. Furthermore, El-Kosary et al. (2013) reported that supplementary irrigation amount of 20 mm applied to peach trees grown under rain fed conditions increase tree yield by 24% in this region.

In the northeastern coast of Egypt, application of 60 mm as a supplementary irrigation for wheat increased grains yield by 75% (Attia and Barsoum 2013). Attalla et al. (2011) reported that application of 20 mm of supplementary irrigation to olive tree increase tree yield by 8%. Similarly, for fig trees, an increase by 11% was found when supplemental irrigation was applied (Allam et al. 2007).

9 Water-Smart Practices to Increase Food Production

9.1 Lands Consolidation

Egypt is characterized by very small land holdings, with 89% of its areas are smaller than 1.3 hectares (FAO 2015). Consolidation of several small farms in the villages to be one large area between 20–30 feddans (8–13 ha) could be done to facilitate the work of agricultural machineries for land preparation, and harvest. It help in unification irrigation dates to save on the applied irrigation water, which will increase water use efficiency by the cultivated crops. It can also improve crops storage and marketing.

9.2 Precise Land Leveling

Precise land leveling is one of these technologies that can be used to reduce the applied water to surface irrigation without any losses in crops productivity, which water use efficiency by the cultivated crops. Previous research indicated that precise land leveling increased water use efficiency for several crops in Egypt. Furthermore, it can save a reasonable amount of the applied irrigation water (Bahnas and Bondok 2010). Other studies indicated that laser land leveling applied to various crops has resulted in water savings up to 15–30% (Jat et al. 2006).

9.3 Raised Beds Cultivation

Basin or furrow cultivation in Egypt is characterized by application of higher irrigation water than other irrigation systems. This results in increasing runoff and deep percolation. On the other hand, cultivation on raised beds proved to reduce the applied water to crops, increase the final yield, and improve rhizosphere environment. Zohry et al. (2020) indicated that the yield of wheat and sunflower was increased by 13 and 11% under raised beds cultivation, compared to the yield resulted under basins cultivation.

9.4 Implementing Intercropping Systems

Intercropping is an agricultural technology that involves planting two or more crop species together, where unit land productivity is increased (harvest two types of crops from the same area) (Porto et al. 2011), water productivity is increased (using the applied amount of irrigation water to one crop to irrigate two crops) and farmers income is increased (reduce risks from crop failure) (Anderson 2007). Intercropping reduces water runoff and soil loss (Lithourgidis 2011), increases nutrient availability (Li et al. 2014), and improves soil quality (Cong et al. 2015). Ouda and Zohry (2018b) indicated that if intercropping systems were included in the national cropping pattern, 2.7 billion cubic meters could be saved, in addition to the increase in total production of the cultivated crops.

9.5 *Improved Water Management Practices of Rain Fed Crops*

Governmental interventions in crop management practices in rain fed areas are now implemented in Egypt to replace the traditional crop production and primitive soil managements with more contemporary management to attain higher yield and to improve soil quality, in addition to improving water harvesting techniques. Farmers were provided with better seeds quality, and advices on fertilizers application.

New studies have advice to implement smart-climate practices in this region by increasing crop diversification through rotating barely or wheat with low water requirements legumes, such as lintel, pea and faba bean to improve the productivity of barley or wheat under rain fed conditions (Khalifa et al. 2004; El-Sadek and Salem 2015). Furthermore, interplanting lintel between young olive, fig and peach trees (young or mature) increased the yield of these trees as a result of increase of soil fertility, reduced wind erosion effects on soil and conserved soil moisture content (Kamel 2011).

Crop rotations, where legume crops followed cereal crops proved to increase the productivity of cereal crops, as these cereal crops benefits from fixed nitrogen by legumes, as well as and the decomposition of legumes roots and nodules (Shams and Kamel 2014). Furthermore, Shams and Kamel (2014) indicated that application of manure is very important because it is a source of copious content of major and minor nutrition elements in available conditions to be absorbed by the roots cereal crops.

9.6 *Simulation of the Effect of Application Supplementary Irrigation in Rain Fed Area*

To determine the effectiveness of the application of supplementary irrigation in the rain fed area, a simulation study was done by Ouda et al. (2016). They simulated the effect of application of 60 mm as a supplementary irrigation to barley or wheat divided into two equal amounts applied one and two months after planting. Their results revealed that barley and wheat yield were increased by 61 and 58%, respectively in the northeastern coast. Similar simulation study was done for olive and peach trees with the application of 60 mm as a supplementary irrigation, which resulted in 42 and 41% increase in their yield, respectively in the same respective region.

Whereas, simulation of the effect of application of supplementary irrigation in the northeastern coast of 60 mm in two doses increased barley and wheat grain yield by 53 and 51%, respectively. Whereas, simulation of the effect of application of supplementary irrigation of 60 mm to olive and fig trees increased yield by 106 and 81% under high drought conditions (Ouda et al. 2016). These results emphasized the important role of application of supplementary irrigation in these rain fed areas.

10 Conclusion

The high population growth rate will put pressure on our water resources in current time and in the future to provide these people with drinking water, which will reduce the amount of water assigned to agriculture. Additionally, the increases in temperature under climate change will lead to greater evaporative losses and will increase the demand on fresh water. Therefore, our only solution is to implement water-smart practices to sustain water resources.

References

Abdel-Shafy HI, Kamel AH (2016) Groundwater in Egypt issue: resources, location, amount, contamination, protection, renewal, future overview. Egypt J Chem 59(3):321–362

Abouelenein R, Oweis T, El Sherif M, Awad H, Khalil FA, Abd El-Hafez S, Karajeh F (2010) A new water saving and yield increase method for growing berseem on raised seed bed in Egypt. Egypt J Appl Sci 25(2A):26–41

Ali MH (2010) Fundamentals of irrigation and on-farm water management, vol 1. Springer Science Business Media, LLC. https://doi.org/10.1007/978-1-4419-6335-2_9

Alkitkat H (2017) Projections of the Egyptian population. In: Future of food gaps in Egypt: obstacles and opportunities. Springer Publishing House. ISBN 978-3-319-46942-3

Allam KhA, Adly MY, Mourad MA (2007) Effect of supplemental irrigation and intercropping treatments on the productivity of fig trees and lentil crop in the Northwest Coast. Misr J Agric Eng 24(1):88–102

Allam MN (2004) Department of Irrigation and Drainage Engineering, Faculty of Engineering, Cairo University: participatory irrigation water management in Egypt: review and analysis, Option Méditerranéennes Series B, n° 48

Anderson RL (2007) Managing weeds with a dualistic approach of prevention and control: a review. Agron Sustain Dev 27:13–18

Ashour MA, El-Attar ST, Amen KA, Abo El-Eyon HM (2010) Irrigation improvement system and its effect on water management in Beni Adi village (case study). J Eng Sci 38(6):1345–1365

Aslam M, Zamir MSI, Afzal I, Yaseen M, Mubeen M, Shoaib A (2013) Drought stress, its effect on maize production and development of drought tolerance through potassium application. Cercetări Agronomice Moldova XLVI 2(154):347–351

Attalla AM, Abdel-Sattar M, Mahrous AE, Abdel-Azeez AA (2011) Olive trees productivity in response to supplemental irrigation under North-Western Coastal conditions in Egypt. Am Eurasian J Agric Environ Sci 11(5):609–615

Attia MA, Barsoum MS (2013) Effect of supplementary irrigation and bio-fertilizer on wheat yield productivity under rain fed conditions. Alex J Agric Res 58(2):149–157

Bahnas OT, Bondok MY (2010) Effect of precision land leveling on faba bean response to compost application in sandy soils. Misr J Agric Eng 27(2):465–481

Batt HA, Merkley G (2009) Water management and user association analysis for irrigation improvement in Egypt. Irrig Drain 59(2):150–160. https://doi.org/10.1002/ird.465

Bauder J (2001) Irrigating with limited water supplies. Montana State University Communications Services, Montana Hall, Bozeman, MT 59717, USA

CAPMAS (2018) Annual statistical year book 2018. Central Agency of Public Mobilization and Statistics

CAPMAS (2019) Annual statistical year book 2019. Central Agency of Public Mobilization and Statistics

Chai Q, Gan Y, Zhao C, Xu H, Waskom RM, Niu Y, Siddique KHM (2016) Regulated deficit irrigation for crop production under drought stress: a review. Agron Sustain Dev 36(3):2–21

Çakir R (2004) Effect of water stress at different development stages on vegetative and reproductive growth of corn. Field Crops Res 89:1–16. https://doi.org/10.1016/j.fcr.2004.01.005

Cong WF, Hoffland E, Li L, Six J, Sun JH, Bao XG (2015) Intercropping enhances soil carbon and nitrogen. Glob Chang Biol 21:1715–1726. https://doi.org/10.1111/gcb.12738

Dakkak A (2016) Egypt's water crisis—recipe for Disaster. https://www.ecomena.org/egypt-water/

Dencic S, Kastori R, Kobiljski B, Duggan B (2000) Evaporation of grain yield and its components in wheat cultivars and land races under near optimal and drought conditions. Euphytica 1:43–52

Dogan E, Kirnak H (2010) Water temperature and system pressure effect on drip lateral properties. Irrig Sci 407–419

Domínguez A, de Juan JA, Tarjuelo JM, Martínez RS, Martínez-Romero A (2012) Determination of optimal regulated deficit irrigation strategies formaize in a semi-arid environment. Agric Water Manag 110:67–77. https://doi.org/10.1016/j.agwat.2012.04.002

El Bedawy R (2014) Water resources management: alarming crisis for Egypt. J Manag Sustain 4(3):108–124

El Quosy DD (2006) A study on matching irrigation supply and demand (MISD). Report prepared for Eco Con Serv-Environmental Solutions. Cairo

El Tahlawi MR, Farrag AA, Ahmed SS (2008) Groundwater of Egypt: "an environmental overview." Environ Geol 55:639–652. https://doi.org/10.1007/s00254-007-1014-1

El-Kosary S, Abdel-Mohsen MA, El-Merghany S, Badran AM (2013) Enhancing the productivity of early grand peaches under Northern Sinai conditions via supplemental irrigation and organic fertilization. J Hortic Sci Ornam Plants 5(2):77–88

El-Sadek A, Salem E (2015) Impact of ridge-farrow water harvesting system on faba bean (Vicia faba L.) production under rain fed conditions in Matrouh, Egypt. Ann Agric Sci 60(1):61–66

Ethan S, Olagoke O, Yunusa A (2016) Effect of deficit irrigation on growth and yield of sugarcane. Direct Res J Agric Food Sci 4(6):122–126

Fabeiro Cortés C, de Santa M, Olalla F, López R, Domínguez A (2003) Production and quality of the sugar beet (Beta vulgaris L.) cultivated under controlled deficit irrigation conditions in a semi-arid climate. Agric Water Manag 62:215–227

Falkenmark M, Lundquist J, Widstrand C (1989) Macro-scale water scarcity requires micro-scale approaches: aspects of vulnerability in semi-arid development. Nat Res Forum 13(4):258–267

FAO (2012) Coping with water scarcity: an action framework for agriculture and food security. FAO Water Report 38. Rome, Italy

FAO (2015) Egypt: wheat sector review. Country highlights. FAO Investment Center. Report No. 21. Rome, Italy

Farkas Z, Varga-László E, Anda A, Veisz O, Varga B (2020) Effects of waterlogging, drought and their combination on yield and water-use efficiency of five Hungarian winter wheat varieties. Water 12:1318

Fereres E, Soriano A (2007) Deficit irrigation for reducing agricultural water use. J Exp Bot 58:147–159

Fernandes-Silva A, Oliveira M, Paço TA, Ferreira I (2018) Deficit irrigation in Mediterranean fruit trees and grapevines: water stress indicators and crop responses. In: Irrigation in agroecosystems. https://doi.org/10.5772/intechopen.80365

Forouzani M, Karami E (2011) Agricultural water poverty index and sustainability. Agro Sustain Dev 31:415–432. https://doi.org/10.1051/agro/2010026

Gan Y, Siddique KHM, Turner NC, Li XG, Niu JY, Yang C, Liu L, Chai Q (2013) Ridge-furrow mulching systems—an innovative technique for boosting crop productivity in semiarid rain-fed environments. Adv Agron 118:429–476. https://doi.org/10.1007/s11104-010-0312-7

Grzesiak S (2001) Genotypic variation between maize (Zea mays L.) single cross hybrids in response to drought stress. Acta Physiol Plant 23:443–456

Gupta NK, Gupta S, Kumar A (2001) Effect of water stress on physiological attributes and their relationship with growth and yield in wheat cultivars at different growth stages. J Agron 86(143):7–1439

Habash DZ, Baudo M, Hindle M, Powers SJ, Defoin-Platel M, Mitchell R et al (2014) Systems responses to progressive water stress in durum wheat. PLoS ONE 9:e108431. https://doi.org/10.1371/journal.pone.0108431

Harbir S, Ingram KT, Jhorar RK (2002) Comparative performance of different water production functions for rice Oryza sativa L. Crop Res Hisar 23:203–213

Heiniger RW (2001) The impact of early drought on corn yield. North Carolina State University. http://www.ces.ncsu.edu/plymouth/cropsci/docs/early_drought_impact_on_corn.html

Hsiao TC, Steduto P, Fereres E (2007) A systematic and quantitative approach to improve water use efficiency in agriculture. Irrig Sci 25:209–231

Hussain S (2005) Effect of supplemental irrigations, seeding rates and foliar application of potassium and macro-micro elements on wheat productivity under rain fed conditions. Sci J Collage Agric 431–454

Inman-Bamber N, Bonnett G, Spillman M, Hewitt M, Jackson J (2008) Increasing sucrose accumulation in sugarcane by manipulating leaf extension and photosynthesis with irrigation. Aust J Agric Res 59:13–26. https://doi.org/10.1071/AR07167

Inman-Bamber N, Lakshmanan P, Park S (2012) Sugarcane for water limited environments: theoretical assessment of suitable traits. Field Crops Res 134:95–104. https://doi.org/10.1016/j.fcr.2012.05.004

Jahan MS, Khanif YM, Sinniah UR (2013) Effects of low water input on rice yield: Fe and Mn bioavailability in soil. Pertanika J Trop Agric 36:27–34

Jat ML, Parvesh C, Raj G, Sharma SK, Gill MA (2006) Laser land leveling: a precursor technology for resource conservation. Rice-wheat consortium technical bulletin Series No. 7. Rice-Wheat Consortium for the Indo-Gangetic Plains, New Delhi

Kamara AY, Menkir A, Badu AB, Ibikunle O (2003) The influence of drought stress on growth, yield and yield components of selected maize genotypes. J Agric Sci 141:43–50

Kamel (2011) Cultural practices to combat degradation under rain fed areas in the Northern Coastal plain in Egypt. Int J Water Resour Arid Environ 1(5):304–311

Keating BA, Herrero M, Carberry PS, Gardner J, Cole MB (2014) Food wedges: framing the global food demand and supply towards 2050. Glob Food Sec 3:125–132

Khalifa HE, Sherif MA, El-Melegy A, Abo Elenien RA, Heliela H, Owies T (2004) Crop rotations and fertilizer type effects on barley production and water-use efficiency at the rain fed area of North Sinai. In: International workshop on management of poor quality water for irrigation; institutional, health and environmental aspects, Moscow, Russia, pp 241–249

Kirigwi F, van Ginkel M, Trethowan R, Sears RG, Rajaram S, Paulsen GM (2004) Evaluation of selection strategies for wheat adaptation across water regimes. Euphytica 135:361–371

Kögler F, Söffker D (2017) Water (stress) models and deficit irrigation: System-theoretical description and causality mapping. Eco Model 361:135–156

Lakshmanan P, Robinson N (2014) Stress physiology: abiotic stresses. In: Moore PH, Botha FC (eds) Sugarcane: physiology, biochemistry, and functional biology. Wiley, Chichester, pp 411–434

Li Q, Bian C, Liu X, Ma C, Liu Q (2015) Winter wheat grain yield and water use efficiency in wide-precision planting pattern under deficit irrigation in North China Plain. Agric Water Manag 153:71–76

Li Y, Zhang W, Ma L, Wu L, Shen J, Davies WJ, Oenema O, Zhang F, Dou Z (2014) An analysis of China's grain production: looking back and looking forward. Food Energy Sec 3:19–32

Lithourgidis AS (2011) Annual intercrops: an alternative pathway for sustainable agriculture. Aust J Crop Sci 5(4):396–410

Liu EK, Mei XR, Yan CR, Gong DZ, Zhang YQ (2016) Effects of water stress on photosynthetic characteristics, dry matter translocation and WUE in two winter wheat genotypes. Agric Water Manag 167:75–85. https://doi.org/10.1016/j.agwat.2015.12.026

Machado R, Ribeiro R, Marchiori P, Machado D, Machado E, Landell M (2009) Biometric and physiological responses to water deficit in sugarcane at different phenological stages. Pesqui Agropecu Bras 44:1575–1582. https://doi.org/10.1590/S0100-204X2009001200003

Ministry of Irrigation and Water Resources (2014) Water scarcity in Egypt: the urgent need for regional cooperation among the Nile Basin Countries. Technical report

Ministry of Water Resources and Irrigation (2005) Arab Republic of Egypt, integrated water resources management plan

Ministry of Water Resources and Irrigation (2006) The national drainage and drainage water reuse programs, Egypt, local actions at the 4th world water forum

Ministry of Water Resources and Irrigation/Planning Sector (2013) Country strategy: national water resources plan Egypt

Mirzaei A, Naseri R, Soleimani R (2011) Response of different growth stages of wheat to moisture tension in a semiarid land. World Appl Sci J 12(1):83–89

Moghazy NH, Kaluarachchi JJ (2020) Sustainable agriculture development in the Western Desert of Egypt: a case study on crop production, profit, and uncertainty in the Siwa Region. Sustainability 12(6568). https://doi.org/10.3390/su12166568

Molden D (ed) (2007) Water for food, water for life: a comprehensive assessment of water management in agriculture. Earth scan/International Water Management Institute, London

Molle F (2019) Egypt. In: Irrigation in the Mediterranean, global issues in water policy, vol 22. https://doi.org/10.1007/978-3-030-03698-0_9

Molle F, Rap E (2014) Brief retrospective on water user organizations in Egypt. Water and salt management in the Nile delta (Project Report No. 3). IWMI, WMRI, Cairo

Morsy M, Sayad T, Ouda S (2017) Present and future water requirements for crops. In: Future of food gaps in Egypt: obstacles and opportunities. Springer Publishing House

Nielsen RL (2007) Assessing effects of drought on corn grain yield. Purdue University, West Lafayette, IN. http://www.kingcorn.org/news/articles.07/Drought-0705.html

Omran ESE (2017) Land and groundwater assessment for agricultural development in the Sinai Peninsula, Egypt. In: Negm AM (ed) The handbook of environmental chemistry. Springer, Berlin, Heidelberg. https://doi.org/10.1007/698_2017_4

Osakabe Y, Arinaga N, Umezawa T, Katsura S, Nagamachi K, Tanaka H et al (2013) Osmotic stress responses and plant growth controlled by potassium transporters in Arabidopsis. Plant Cell 25:609–624. https://doi.org/10.1105/tpc.112.105700

Ouda S, Zohry A (2018a) Water requirements for prevailing cropping pattern. In: Cropping pattern to overcome abiotic stresses: water, salinity and climate. Springer Publishing House

Ouda S, Zohry A (2018b) Cropping pattern to face water scarcity. In: Cropping pattern to overcome abiotic stresses: water, salinity and climate. Springer Publishing House

Ouda S, Noreldin T, Amer A (2016) Rain fed areas in Egypt: obstacles and opportunities. In: Management of climate induced drought and water scarcity in Egypt: unconventional solutions. Springer Publishing House. ISBN 978-3-319-33659-6

Ouda S, Zohry A, Noreldin T (2020) Water scarcity leads to food insecurity. In: Deficit irrigation: a remedy for water scarcity. Springer Publishing House. ISBN 978-3-030-35585-2

Pandit M, Chakravarty MK, Chakraborty M, Tudu VK, Sah RP, Narayan SC (2017) Identifcation of maize genotypes for moisture stress tolerance. Int J Basic Appl Biol 4(2):74–81. https://doi.org/10.1155/2018/3158474

Pandit M, Sah RP, Chakraborty M, Prasad K, Chakraborty MK, Tudu V, Narayan SC, Kumar A, Manjunatha N, Kumar A, Rana M (2018) Gene action and combining ability for dual purpose traits in maize (Zea mays L.) under water deficit stress prevailing in eastern India. Range Mgmt Agrofor 39(1):29–37. https://doi.org/10.1080/15427528.2017.1315625.

Pantuwan G, Fukai S, Cooper M, Rajatasereekul S, O'Toole JC (2000) Field screening for drought resistance. In: ACIAR Proceedings No.101. Increased lowland rice production in the Mekong Region: Proceedings of an international workshop held in Vientiane, Laos, 30 October–2 November 2000, pp 69–77

Porto VCN, Bezerra NF, Lima JSS, Barros JAP, Moreira JN (2011) Combination of lettuce and rocket cultivars in two cultures intercropped with carrots. Horti Bras 29:404–411

Raskin P, Gleick P, Pontius G, Strzepek K (1997) Water futures: assessment of long-range patterns and prospects. Stockholm Environment Institute, Stockholm

Ravindra K, Tedia K, Malaiya S, Yerne A (2002) Effect of drought on root and shoot growth, plant water status, canopy temperature and yield of rice. J Soils Crops 12:179–182

Sadras VO, Villalobos FJ, Orgaz F, Fereres E (2016) Effects of water stress on crop production. In: Villalobos F, Fereres E (eds) Principles of agronomy for sustainable agriculture. Springer, Cham. https://doi.org/10.1007/978-3-319-46116-8_14

Sah M, Chakraborty K, Prasad M, Pandit VK, Tudu MK, Chakravarty SC, Narayan MR, Moharana RP (2020) Impact of water deficit stress in maize: phenology and yield components. Sci Rep 10:2944. https://doi.org/10.1038/s41598-020-59689-7

Sepaskhah AR, Ghahraman B (2004) The effects of irrigation efficiency and uniformity coefficient on relative yield and profit for deficit irrigation. Biosys Eng 87(4):495–507

Shams AS, Kamel AS (2014) Rotations in Coastal plains to combat desertification in Egypt. Int J Water Resour Arid Environ 3(2):121–131

Singh S, Rao PNG (1987) Varietal differences in growth characteristics in sugarcane. J Agric Sci 108:245–247

Smit M, Singels A (2006) The response of sugarcane canopy development to water stress. Field Crops Res 98:91–97. https://doi.org/10.1016/j.fcr.2005.12.009

Sojka RE, Bjorneberg DL, Strelkoff TS (2007) Irrigation-induced erosion. American Society of Agronomy, Crop Science Society of America, Soil Science Society of America, 677 S. Segoe Rd., Madison, WI 53711, USA, Irrigation off-Agricultural Crops. 2nd ed., Agronomy Monograph no 30

Taha A, Ouda S (2016) Deficit irrigation for wheat and maize grown in sandy soil to face water scarcity. In: 4th African regional ICID conference, 24–28 April, Aswan, Egypt

Tariq JA, Usman K (2009) Regulated deficit irrigation scheduling of maize crop. 2009. Sarhad J Agric 25(3):441–450

The World Bank (2020) Modernizing irrigation improved water security for farmers in Egypt. https://www.worldbank.org/en/results/2020/04/01/modernizing-irrigation-improved-water-security-for-farmers-in-egypt

UNEP (2014) Green economy scoping study: Egypt. https://www.greengrowthknowledge.org/sites/default/files/downloads/resource/Green_Economy_Scoping_Study_Egypt_UNEP.pdf

United Nations (2006) Coping with water scarcity: a strategic issue and priority for system-wide action. UN-Water Thematic Initiatives August 2006

Vlek PLG, Hillel D, Braimoh AK (2008) Land use and soil resources. Springer Science + Business Media BV

Yang X, Wang B, Chen L, Li P, Cao C (2019) The different influences of drought stress at the flowering stage on rice physiological traits, grain yield, and quality. Sci Rep 9:3742

Zareian A, Abad HHS, Hamidi A (2014) Yield, yield components and some physiological traits of three wheat (Triticum aestivum L.) cultivars under drought stress and potassium foliar application treatments. Int J Biosci 4(5):168–175

Zohry A, Ouda S (2020) egypt faces water deficiency and food insufficiency. In: Deficit irrigation: a remedy for water scarcity. Springer Publishing House
Zohry A, Ouda S, Abdel-Wahab T (2020) Sustainable intensive cropping to reduce irrigation-induced erosion: intercropping systems under surface irrigation practice. Moroccan J Agric Sci 1(2):63–71

Chapter 2
Soil-Smart Practices: Integrated Soil Fertility Management

Abd El-Hafeez Zohry and Samiha Ouda

Abstract In this chapter, we discussed several aspects related to soil-smart practices. Application of integrated fertilizers, where chemical fertilizers were partially replaced by biofertilizers, compost and/or manure was overviewed. Moreover, we compared between application of chemical fertilizer versus nonchemical fertilizer, or integrated chemical and nonchemical fertilizer to three edible oil crops in Egypt, namely soybean, sunflower and canola cultivated in the sandy soil of East El-Owinat region and their effects on attaining high yield. The results showed that the three crops followed the same pattern, where application of 100% chemical fertilizer only attained the highest yield and application of 50% compost and 50% biofertilizer attained the highest yield under integrated fertilizer practice. Thus, the sustainable use of soils necessitates the reduction of the use of chemical fertilizers to reduce soil and groundwater pollutions. Additionally, application of the integrated fertilizer can improve soil fertility and obtain healthy seeds as organic seeds, which attain higher prices compensate for yield losses.

Keywords Biofertilizer · Cow manure · Compost · Chemical fertilizer · Integrated fertilizer sources · Soybean · Sunflower · Canola

1 Introduction

Soil fertility management is the most important factor in the sustainable crop production that can attain the first pillar of climate-smart agriculture. Vanlauwe et al. (2010) defined integrated soil fertility management as "a set of soil fertility management practices that necessarily include the use of fertilizer, organic inputs, and improved germplasm combined with the knowledge on how to adapt these practices to local

A. E.-H. Zohry (✉)
Crops Intesefication Research Department, Field Crops Research Institute; Agricultural Research Center, Giza, Egypt

S. Ouda
Water Requirements and Field Irrigation Research Department, Soils, Water and Environment Research Institute; Agricultural Research Center, Giza, Egypt

S. Ouda and A. E.-H. Zohry, *Climate-Smart Agriculture*,
https://doi.org/10.1007/978-3-030-93111-7_2

conditions, aiming at maximizing agronomic use efficiency of the applied nutrients and improving crop productivity".

Many studies reported yield increases and benefits arising from application of integrated soil fertility management interventions, which is mainly attributed to improving nutrient use efficiency of organic or inorganic fertilizers (Serafim et al. 2013). The higher utilization efficiency either from agronomic and/or nutrient recovery efficiency obtained from integrated soil fertility management (as compared to sole application) may be attributed to the principles of synergistic effects involving different mechanisms (Hai-cheng et al. 2018). For example, combinations of mineral fertilizers and leguminous trees have been used to improve soil fertility in an agroforestry ecosystem (Palm et al. 2001). Legumes play a significant role in sustainable agriculture through their ability to improve soil fertility, it can improve nitrogen amount through biological N-fixation (Hai-cheng et al. 2018). Additionally, to maximize such functions, legumes need more phosphorus for energy transformation in nodules (Oldroyd and Dixon 2014).

The combined use of mineral and organic manure has been used to improve soil fertility and yield of different crops (Abiven et al. 2009). Evidences shows that the use of organic fertilizers combined with inorganic sources enhances nutrient availability, and improves crop productivity (Usman et al. 2015), as well as improves environmental conditions and public health (Kombiok et al. 2008). Combining inorganic and organic resources resulted in high improvements in soil fertility and crop growth (Endris and Dawid 2015). Furthermore, integration of composts in integrated soil fertility management systems with inorganic fertilizer source was also reported to enhance crops growth and yield (Zingore et al. 2015).

In this chapter, we discussed several aspects related to the first pillar of climate-smart agriculture, namely soil-smart practices. Application of integrated fertilizers, where chemical fertilizers were replaced with biofertilizers, compost and manure was also discussed and a case study was presented.

2 Fertilizer Application

Application of fertilizer is an important element in increasing the yield of crops. However, there are large increases in the use of chemical fertilizer, which resulted in leaching it by the runoff and an increase of its concentration in the groundwater (Wu and Ma 2015). The continual use of chemical fertilizers in farming is known to affect soil pH of the farms leading to acidic environment that promotes nutrients solubility, as well as bioavailability and mobility of heavy metals (Elliott et al. 2013).

Fertilizers are grouped as chemical fertilizer (nitrogen, phosphorous and potassium fertilizers) and nonchemical fertilizer, namely organic (farmyard manure and compost) and biofertilizers (Khan et al. 2018).

2.1 Chemical Fertilizer

2.1.1 Nitrogen

Nitrogen is the main nutrient affecting crop growth and yield formation, nevertheless the excessive amount of nitrogen in relation to the crop requirements has considerable negative environmental impacts (Sutton et al. 2011). Korkmaz (2007) stated that only 50% of the applied chemical nitrogen is used by the plants, 2–20% is lost through evaporation and 15–25% reacts with organic components in the soil and the remaining penetrates to the ground water. Thus, effective management of nitrogen fertilizer is essential for optimizing crop yields, maximizing farmer income and minimizing negative environmental impacts from reactive nitrogen flow to groundwater (Doltra et al. 2011). There is a challenge associated with intensive agricultural production systems, namely to combine intensive production with high nitrogen use efficiency to attain food security and minimized negative environmental impacts. This includes strategies of chemical nitrogen application with other source of nitrogen (Sutton et al. 2011).

2.1.2 Phosphorus

Phosphorus plays a significant role in root development and nutrients uptake. Similar to nitrogen, its use resulted in unprecedented increases in yield (Blessing et al. 2017). Phosphorus in the soil exists as a heterogeneous mixture of adsorbed, absorbed (diffused into the solid or embedded in the lattice structure), and mineral species, each on their own spectrum of solubility; some associations are easily reversible, while others are not (Barrow 2015). Permanent phosphorus losses from the field include leaching into tile drainage, soluble phosphorus in runoff, and erosion of soil particles containing plosphorus (Simpson et al. 2011).

2.1.3 Potassium

Potassium is a vital macronutrient for plants, required for normal plant growth and development (Wakeel et al. 2011). It can be fix within the soil depending upon the charge of clay minerals; which reduce its availability to the crops (Hussain et al. 2011). It is an essential nutrient and its availability controls many biochemical and physiological processes in plants such as enzyme activation, photosynthesis, protein synthesis, osmoregulation, energy transfer, stomatal movement, cation–anion balance and stress resistance (Khan and Inamullah 2014). Potassium is also involved in many metabolic pathways and affects crop quality hence it is often called "the quality element" (Ali et al. 2019).

2.2 Nonchemical Fertilizer

2.2.1 Organic Fertilizer

Organic fertilizer is an alternative fertilization method to avoid soil degradation consequences (Cvijanovic et al. 2012). They are naturally occurring compounds produced from farm by-products, which assess in the sustainable use of available resources (Mondini and Sequi 2008). It can partially substitute for chemical nitrogen, phosphorus and potassium. It is characterized by slow release of the nutrients in the soil (Raj and Vasanthi 2019). It has many benefits, where its low amounts of nitrogen will be immediately absorbed by the growing plants and that will leave no nitrogen available for weeds to uptake in the rhizosphere (Mohammadi and Rokhzadi 2012). Using organic fertilizer led to an increase in soil organic matter contents, improves soil structure and enhance root growth (Mohammadi and Rokhzadi 2012), it maintains soil aeration, and growth of soil microorganisms (Jenkinson 1988) and it improves water holding capacity (Posner et al. 2008). Using organic fertilizer reduces the attraction rate of pests to the growing plants as a result of free nitrogen in plant tissue (Mohammadi and Rokhzadi 2012). Additionally, it minimizes problems related to farm wastes disposal and reduces environmental pollution caused by excessive application of mineral fertilizers (Maiorana et al. 2005).

Organic agriculture is a system of ecological production management, which promotes and improves biodiversity, matter circulation and biological soil activity (Cvijanovic et al. 2010). It involve using organic fertilizer in crop production. However, organic crops production are characterized by relatively low yield, compared to conventional crop production using chemical fertilizer. The low production can be compensated by high price of the organic products (Hartman et al. 2016).

Farmyard manure is a type of organic fertilizer, which is considered as a valuable input to soil for higher crop production (Ezung and Jami 2019). It improves soil health by inducing favorable physical, chemical and biological conditions (Alam et al. 2010), it has a high organic matter, and nitrogen contents (Moral et al. 2005) and a variety of micronutrients (Cayuela et al. 2009). It is a cheap way for a society to conserve the environment (El Sheikha 2016).

Compost is another type of organic fertilizer and an alternative soil amendment, which characterized by its slower nutrient release potential, compare to chemical fertilizer (Akanbi et al. 2007). Furthermore, adding compost improves soil fertility by increasing both the quantity and the quality of soil organic matter (Rivero et al. 2004). According to Eghball (2002), compost application in excess of crop requirements can last for several years in the soil since only a fraction of nitrogen and other nutrients in compost become available in the first year after application.

2.3 Biofertilizer

Biofertilizers are eco-friendly and have been proved to be effective and economical alternate of chemical fertilizers with lesser input of capital and energy (Hafeez et al. 2002). It is an important part of environmental friendly sustainable agricultural practices (Akbari et al. 2011). It refers to living microorganisms in a symbiotic and asymbiotic way to supply nutrients to the growing plants, resulting in significant improvement in crops yield and reduction of soil nutrients depletion (Dhanasekar and Dhandapani 2012). It can decrease soil pH, which led to an increase in the availability of trace elements that enhance plant growth (Hafeez et al. 2002). It helps in increasing gibberellin level, which increases root growth, where higher rate of root growth increases the uptake of soil nutrients, increases the permeability of the soil, and increases soil water holding capacity (Hamidi et al. 2015).

3 Integration of Different Fertilizer Sources

There is a recent emphasis on the integrated use of different sources of nutrients, such as organic manure, compost and biofertilizer in combination with chemical fertilizers for its known benefits (Patidar and Malhi 2004). These integrations can enhance integrated soil fertility management and attain the objectives of climate-smart practices.

4 Integration Fertilizer Management for Edible Oil Crops: Evidence from Egypt

There is a large gap between the production and consumption of edible oil in Egypt. In 2017, Egypt production of the oil crops was 380 thousand tons, where soybean and sunflower production contributed by 45 and 19 thousand tons, respectively. On the other hand, the total importation of edible oil was 591 thousand tons, of them 461 and 73 thousand tons were for soybean and sunflower, respectively (CAPMAS 2019).

The three important crops that can contribute in reducing edible oil gap in Egypt are soybean, sunflower and canola. Soybean seeds contain 43% protein, 20% fat, 21% carbohydrate and other nutrients (calcium, phosphorus, and iron), as well as vitamins (Rahman 1982). It is a pulse crop with multiple food and economic advantages. It has the capacity to fix atmospheric nitrogen from the air through symbiosis with rhizobium bacteria and thus contributes to improved soil fertility (Dianda et al. 2016). Whereas, sunflower oil contains large amount of vitamins (A, D, E and K) and 20–40% proteins (Dhanasekar and Dhandapani 2012). The importance of sunflower is relay on its high oil quality, as well as it contains large amount of vitamins

(Dhanasekar and Dhandapani 2012). It is adapted to warm season and it is considered drought tolerant crop (Ahmed et al. 2015). Furthermore, previous research in Egypt indicated that canola could be successfully grown in newly reclaimed lands in order to avoid competition with other winter crops grown in the old cultivated area (Sharaan and Gallab 2002). Canola seeds contain 40–45% oil and 36–40% protein. Canola oil has low content of erucic acid and glucosinolates, and it has high content of omega 3 and vitamin E (Tomm et al. 2010).

Several investigations were done on the effect of integration of organic fertilizer with chemical fertilizer locally and internationally. For example, Khamparia, et al. (2018) indicated that integration of chemical fertilizer with organic fertilizer positively affected soil health and soybean yield. Khaim et al. (2013) indicated that application of chemical fertilizer in combination with biofertilizer to soybean increased its yield and yield contributing characters. Similar trend was obtained when chemical fertilizer in combination with biofertilizer was applied to sunflower (Dhanasekar and Dhandapani 2012). Furthermore, Mohamed et al. (2017) indicated that the interaction among different levels of mineral nitrogen fertilizer and biofertilizer attained higher canola yield, compared to the application of either fertilizer source.

5 Case Study of East El-Owinate Region

One of the solution to reduce the of edible oil gap is to cultivate it in the newly reclaimed areas outside the Nile Delta and Valley. A promising region in Egypt, namely East El-Owinate may provide an opportunity to overcome some of deficiency in edible oil production. It could also help in reducing competition exists in the old lands between oil crops and other strategic crops. The appropriate combination of chemical and nonchemical fertilizers (biofertilizer, compost and cow manure) are very important determinant in the production of these crops to attain high yield, and to encourage organic cultivation in this version soil of East El-Owinat region.

To attain that, a study was implemented in East El-Owinat (latitude = 22.3 N, longitude = 28.0 E and elevation = 128.3 m above sea level) by Zohry and Mohamadin (2020) during the two growing seasons of 2016/17 and 2017/18 in sandy soil under sprinkler system. The soil physical analysis revealed that the percentage of clay, silt and sand were 3, 8 and 89%, respectively. The chemical analysis of the soil revealed that soil pH was 8.5, EC was 0.1 dS/m and $CaCo_3$ was 4.0. Maximum temperature in the winter season was between 22–33 °C and in the summer season is between 36–38 °C. Whereas, minimum temperature in the winter season was between 4–16 °C and in the summer season was between 20–21 °C.

Seven fertilizer treatments were applied to the three studied crops: control (100% chemical fertilizer, 100%CF); nonchemical fertilizer treatments (100% biofertilizer, (100%BF); 100% cow manure (100%CM); and 100% compost (100%CP)); and integrated fertilizer sources (50% chemical fertilizer and 50% biofertilizer (50%CF + 50%BF); 50% cow manure and 50% biofertilizer (50%CM + 50%BF); and 50% compost and 50% biofertilizer (50%CP + 50%BF)).

For the chemical fertilizer treatment (control), the recommended dose of fertilizer in sandy soil for the three crops was obtained from Legume Crops Department and the recommended dose for sunflower and canola was obtained from Oil Crops Department; Field Crops Research Institute; Agricultural Research Center; Egypt.

For the three crops, nitrogen fertilizer in the rate of 144 kg N/ha was added in three doses, the first dose at planting date (20% in the form of ammonium sulphate (20.6% N). The second and the third doses were added as 40% each of the total dose in the form of ammonium nitrate (33.5% N), 30 and 45 days after sowing, respectively. Calcium super phosphate (15.5% P_2O_5) was added at the rate of 74.4 kg P_2O_5/ha during land preparation and potassium sulphate (48% K_2O) was added at the rate of 108.2 kg K_2O/ha, 30 days after planting. The rest of the studied treatments were adjusted to contain the recommended nitrogen fertilizer dose for each crop.

With respect to the nonchemical fertilizer treatments, biofertilizer treatment was applied using Micropbin (Azospirilum sp., Azotobacter sp., Bacillus megatheruim var. phosphaticum, Pseudomonas sp., and Mycorriza sp.) obtained from the General Organization of Agricultural Equalization Fund (GOAEF), Ministry of Agriculture and Land Reclamation, Egypt. The recommended dose of inculcation with the biofertilizer was performed by mixing 1 g of bacteria with 100 g of seeds of each crop. Thus, each seed received approximately 10 million bacteria on its surface. Inculcation with biofertilizer treatment was performed by mixing it with the seeds before sowing according to seeding rate of each crop.

Old cow manure was obtained from cow farm near the experimental site. It was added for each crop before land ploughing. The chemical composition of the used cow manure and manure are presented in Table 1. The used compost was ready-made manufactured using rice hay (60%), farmyard manure (25%), poultry manure (10%) and fertile soil (5%).

To summarize, the amount of nonchemical and chemical nitrogen fertilizer applied to the three crops, which represent the recommended dose of nitrogen fertilizer are presented in Table 2.

No other fertilizers were applied to the three crops under biofertilizer, cow manure or compost treatments.

Table 1 Chemical composition of cow manure and compost used in the experiment

Character	Cow manure	Compost
Organic matter (%)	13.6	24.6
Total nitrogen (%)	0.35	0.41
Total phosphorus (%)	0.25	0.28
Total potassium (%)	0.10	0.31

Table 2 Amount of different source of nitrogen fertilizer applied to each crop

Crop	Biofertilizer (g)	Cow manure (m^3/ha)	Compost (ton/ha)	Chemical (kg N/ha)
Soybean	840	63.4	24	144
Sunflower	120	63.4	24	144
Canola	60	63.4	24	144

Table 3 Soybean, sunflower and canola yield as affected by chemical fertilize in 2016/17 and 2017/18 growing seasons

Treatments	Growing season	Yield (ton/ha)
Soybean yield	2016	2.71
	2017	2.74
Sunflower yield	2016	3.10
	2017	3.12
Canola yield	2016/17	2.87
	2017/18	2.97

6 Effect of Application of Chemical Fertilizer

The results in Table 3 clearly showed that the yield of the three crops were significantly affected by fertilizer treatments in both growing seasons. The results also showed that the application of 100% chemical fertilizer attained high yield of the three crops, which could be attributed to the fact that chemical fertilizer is easily accessible to plants roots in the rhizosphere (Khan et al. 2018). Furthermore, Ahmadi and Bahrani (2009) indicated that canola seed yield were positively correlated to soil nitrogen level. However, on the long run, the continuous application and excessive use of chemical fertilizer will cause environmental problems. Lal (2011) indicated that, in developing countries, most of the farmers facing increasing fertilizer costs, environmental degradation and food quality deterioration due to the unnecessary application of chemical fertilizer.

7 Effect of Application of Nonchemical Fertilizer

Table 4 showed that, under the nonchemical fertilizer treatments, the lowest values of the yield of the three crops were obtained under application of 100% of biofertilizer. Although, Hayat et al. (2010) indicated that Azotobacter and Azospirillum has positive direct and indirect effects on plant growth and pest resistance. The results of this experiment showed that application of biofertilizer produced low yield values, compared to chemical fertilizer treatments. Billah and Bano (2015) indicated that the availability of organic matter in the rhizosphere is a proponent for favorable conditions and it affects plant–microbe interactions. That statement explains the low yield

Table 4 Soybean, sunflower and canola yield as affected by nonchemical fertilize in 2016/17 and 2017/18 growing seasons

Treatments	Soybean yield (ton/ha)		Sunflower yield (ton/ha)		Canola yield (ton/ha)	
	2016	2017	2016	2017	2015/16	2016/17
100%BF	0.82	0.77	1.27	1.80	0.96	1.14
100%CM	1.42	1.46	1.73	2.16	1.19	1.43
100%CP	1.66	1.63	1.75	2.40	1.79	1.92

BF = biofertilizer, CM = cow manure, CP = compost

values of the three crops under application of biofertilizer only, where the version soil of the experiment having low organic content.

The highest yield of the three crops were obtained under the application of 100% compost. Soybean yield under application of 100% compost was higher by 51 and 53% in the first and second season, respectively, compared to application of 100% of biofertilizer (Table 4). Hammed et al. (2019) indicated that organic fertilizer with animal-based had less positive effect on soybean, compared to organic fertilizer with plant-based. Furthermore, the table also indicated that application of 100% compost, resulted in the highest sunflower yield by 38 and 33%, compare to application of 100% biofertilizer. Similar results were obtained by Akbari et al. (2011), when they individually applied biofertilizer and compost to sunflower. Mohammadi and Rokhzadi (2012) stated that compost have more positive effect in stimulating soil microbial activity than organic manure. Whereas, Ali et al. (2011) and El Sabagh et al. (2015) stated that the individual application of biofertilizer, compost and organic manure increased canola yield, compared to no nitrogen application.

Although yield of the three crops was lower under the application of nonchemical fertilizer, the higher price attained from selling the yield as an organic product could compensate the loss in the productivity. Hartman et al. (2016) indicated that the price of organic seeds almost three folds of the price of seeds produced using chemical fertilizers.

8 Effect of Application of Integrated Fertilizer

The results in Table 5 showed that integrated fertilizer treatments attained higher yields, compared to application of each one of nonchemical treatments (Table 4).

Table 5 Soybean, sunflower and canola yield as affected by integrated fertilizer in 2016/17 and 2017/18 growing seasons

Treatments	Soybean yield (ton/ha)		Sunflower yield (ton/ha)		Canola yield (ton/ha)	
	2016/17	2017/18	2016/17	2017/18	2015/16	2016/17
50%CF + 50%BF	2.33	2.28	2.27	2.88	2.40	2.63
50%CM + 50%BF	1.46	1.63	1.88	2.64	1.44	1.61
50%CP + 50%BF	1.75	1.82	1.94	2.64	2.30	2.39

CF = chemical fertilizer, BF = biofertilizer, CM = cow manure, CP = compost

8.1 *Application of 50% Chemical Fertilizer and 50% Biofertilizer*

Application of 50% chemical fertilizer and 50% biofertilizer treatment attained the highest values of the yield of the three crops under the integrated fertilizer treatments. Soybean yield was increased by 37 and 36% under application of 50% chemical fertilizer and 50% biofertilizer, in the first and second season, respectively (Table 5). It was reported by Abou Elkhair et al. (2014), Attia et al. (2014) and Khaim et al. (2013) that the application of biofertilizer with chemical fertilizer to soybean resulted in a significant increase in its yield.

It was also noticed from the table that soybean yield was lower in the second season, compared to the first season. Zerihun et al. (2013) indicated that cultivation of soybean as a legume crop is expected to have positive effects on soil fertility because it fix nitrogen, in addition to its residue decomposition and that should benefit the following crop. However, in this experiment, wheat followed soybean in the first growing season. It seemed that cultivation of wheat depleted all the nutrients left in the soil after soybean cultivation in such low fertile soil, which negatively affected soybean yield in the second growing season. Woźniak and Soroka (2018) indicated that cultivation of wheat could involve adverse changes in the soil including decreasing contents of organic matter.

With respect to sunflower, Arif et al. (2016) stated that the application of chemical nitrogen fertilizer with biofertilizer positively affected sunflower yield, as well as soil chemical and biological fertility. Similar trends were reported by other authors, where Khandekar et al. (2018) indicated that sunflower yield was improved under the combination of nitrogen fertilizer and biofertilizer, compared to using either fertilizer type alone. Furthermore, Namvar et al. (2012) stated that the highest grain yield were obtained from the highest level of nitrogen fertilizer (200 kg N/ha) and biofertilizer inoculation. Khandekar et al. (2018) reported that sunflower yield was improved under the combination of nitrogen fertilizer and biofertilizer compared to using either fertilizer type alone. It could be also noticed from Table 5 that that sunflower yield was higher in the second season, compared to the first season because it was preceded by faba bean in the second growing season.

Similarly, the same trend was obtained for canola by Ali et al. (2011) and Mohamed et al. (2017), where they indicated that the interaction among the used levels of mineral fertilizer and biofertilizer attained higher canola yield, compared to no fertilizer application. Moreover, application of 50% compost and 50% biofertilizer treatment attained lower yield than the application of 50% chemical fertilizer and 50% biofertilizer treatment by 4 and 9% in the first and second seasons, respectively. The results in Table 5 also showed that canola yield was higher in the second growing season, compared to the first growing season as a result of peanut preceded canola in the second growing season.

8.2 *Application of 50% Cow Manure and 50% Biofertilizer*

The results in Table 5 showed that integrated fertilizer treatments attained higher yields, compared to application of one nonchemical treatment. Application of 50% cow manure and 50% biofertilizer attained the lowest yield of the three crops in the integrated fertilizer treatments. Furthermore, it could be noticed from the table that soybean yield was improved in the second year under the rest of the studied treatments, which can be attributed to the residual effect of the organic fertilizer treatments. In this context, Khaim et al. (2013) indicated that the highest soybean yield was found when 100% of chemical fertilizer was applied, which was statistically identical with the application of 75% of chemical fertilizer and 1 ton/ha of poultry manure, as well as the application of 75% of chemical fertilizer and 3 ton/ha cow manure. The reduction in chemical fertilizer cost could occurred under application of 50% compost and 50% biofertilizer treatment, in addition to higher price of soybean as an organic product could be gain. Hartman et al. (2016) indicated that the price of organic soybean seed almost three folds of the price of seeds produced using chemical fertilizers.

With respect to canola, Abd El-Gawad et al. (2009) indicated that application of biofertilizer with organic manure resulted in high improvement in canola yield attributed components and yield. Whereas, application of sheep manure with biofertilizer increased canola plant height, number of branches/plant, and seed yield (Abd El-Gawad et al. 2009).

8.3 *Application of 50% Compost and 50% Biofertilizer*

Furthermore, application of 50% compost and 50% biofertilizer achieved the second best yield of the three crops. For soybean, the yield was lower by 25 and 20% in the first and second season, respectively, compared to the yield obtained under 50% chemical fertilizer + 50% biofertilizer treatment (Table 5). Similarly, for sunflower, the yield was lower than the yield resulted from application of 50% chemical fertilizer and 50% biofertilizer by 15 and 8% in the first and second seasons, respectively.

Furthermore, there were no difference between application of 50% cow manure and 50% biofertilizer treatment and the application of 50% compost and 50% biofertilizer treatments, where the yield under the first treatment was lower by 3 and 0% than its value in the second treatment in the first and second season, respectively (Table 5). It could be also noticed from the table that sunflower yield was higher in the second season, compared to the first season because it was preceded by faba bean in the second growing season.

Similar trend was obtained for canola, where application of 50% compost and 50% biofertilizer treatment attained lower yield than the application of 50% chemical fertilizer and 50% biofertilizer treatment by 4 and 9% in the first and second seasons, respectively (Table 5). Several researchers studied the application of integrated fertilizer to canola. Kazemeini et al. (2010) reported that 50% of the required nitrogen fertilizer to canola could be replaced by compost application. El Sabagh et al. (2015) indicated that application of biofertilizer could increase the efficiency of nitrogen uptake and it could be more effective with the application of compost. Abd El-Gawad et al. (2009) indicated that application of biofertilizer with organic manure resulted in high improvement in canola yield attributed components and yield. Kazemeini et al. (2010) reported that 50% of the required nitrogen fertilizer to canola could be replaced by compost application. Furthermore, the results in Table 5 also showed that canola yield was higher in the second growing season, compared to the first growing season as a result of peanut preceded canola in the second growing season.

9 Conclusion

The sustainable use of soils and soil-smart practices necessitates the reduction or elimination of the use of chemical fertilizers to reduce soil and groundwater pollutions. The appropriate combination of integrated fertilizer sources to be applied to soybean, sunflower and canola cultivated in the sandy soil of East El-Owinat region were studied to attain high yield, and to encourage organic cultivation in this new areas. The results showed that the three crops followed the same trend, where application of non-chemical fertilizer, namely 100% compost attained the highest yield, compared to the nonchemical fertilizer. Furthermore, application of integrated fertilizer, namely application of 50% compost in addition to 50% biofertilizer attained the higher yield than application of nonchemical fertilizer. Thus, it could be concluded that application of the integrated fertilizer increase the yield of the three studied crops, in addition it could improve soil fertility and obtain healthy seeds as organic seeds with higher prices and, consequently compensate for yield losses.

References

Abd El-Gawad AM, Hendawey MH, Farag HIA (2009) Interaction between biofertilization and canola genotypes in relation to some biochemical constituents under Siwa Oasis conditions. Res J Agric Biol Sci 5(1):82–96

Abiven S, Menasseri S, Chenu C (2009) The effects of organic inputs over time on soil aggregate stability—a literature analysis. Soil Biol Biochem 41:1–12

Abou Elkhair WA, El-Shaboury MGH, Abd El-Aal HT (2014) Influence of multifunctional bio-formulations on productivity of some soybean varieties. Middle East J Agric Res 3(4):1083–1089

Ahmadi M, Bahrani MJ (2009) Yield and yield components of rapeseed as influenced by water stress at different growth stages and nitrogen levels. Am-Euras J Agric Environ Sci 5(6):755–761

Ahmed MH, Mohamed SM, Moohamed HM, Shahata HM (2015) Effect of bio, nitrogen and phosphorus fertilizers on growth, yield and yield components of sunflower crop grown in El-Kharga Oasis, New Valley. Assiut J Agric Sci 46(2):148–161

Akanbi WB, Adebayo TA, Togun OA, Adeyeye AS, Olaniran OA (2007) The use of compost extract as foliar spray nutrient source and botanical insecticide in Telfairia occidentalis. World J Agric Sci 3(5):642–652

Akbari P, Ghalavand A, Modarres SAM, Agha AM (2011) The effect of biofertilizers, nitrogen fertilizer and farmyard manure on grain yield and seed quality of sunflower (*Helianthus annus L.*). J Agric Technol 7(1):173–184

Alam M, Siddiqua A, Chowdhury M, Prodhan M (2010) Nodulation, yield and quality of soybean as influenced by integrated nutrient management. J Bangladesh Agric Univ 7:229–234

Ali S, Hafeez A, Ma X, Tung SA, Chattha MS, Shah AN, Yang G (2019) Equal potassium-nitrogen ratio regulated the nitrogen metabolism and yield of high-density late-planted cotton (Gossypium hirsutum L.) in Yangtze River valley of China. Ind Crop Prod 129:231–241

Ali ME, Fathi AI, Mohamed OH, El-Edfawy YM (2011) Response of canola productivity and quality to bio-organic and inorganic Nfertilizers. J Soil Sci Agric Eng Mansoura Univ 2(12):1255–1272

Arif SM, Riaz M, Shahzad MS, Yasmeen T, Akhtar JM, Raiz AM, Jassey EJV, Bragazza L, Buttler A (2016) Associative interplay of plant growth promoting rhizobacteria (*Pseudomonas aeruginosa* QS40) with nitrogen fertilizers improves sunflower (*Helianthus anniis L.)* producivity and fertility of aridisol. Appl Soil Ecol 108:238–247

Attia AM, El-Guibal HA, Shaban KhA, Abdel Mhosen MI (2014) Influence of applied biofertilizer on productivity, quality and nutrients content of some soybean cultivars under saline soil conditions. J Soil Sci Agric Eng Mansoura Univ 5(12):1647–1666

Barrow NJ (2015) Soil phosphate chemistry and the P-sparing effect of previous phosphate applications. Plant Soil 397:401–409. https://doi.org/10.1007/s11104-015-2514-5

Billah M, Bano A (2015) Role of plant growth promoting rhizobacteria in modulating the efficiency of poultry litter composting with rock phosphate and its effect on growth and yield of wheat. Waste Manag Res 33:63–72

Blessing OC, Ibrahim A, Safo EY, Yeboah E, Abaidoo RC, Logah V, Monica UI (2017) Fertilizer micro-dosing in West African low-input cereals cropping: benefits, challenges and improvement strategies. Afr J Agric Res 12:1169–1176

CAPMAS (2019) Central agency of public mobilization and statistics. Annual statistical year book 2019

Cayuela ML, Sinicco T, Mondini C (2009) Mineralization dynamics and biochemical properties during initial decomposition of plant and animal residues in soil. Appl Soil Ecol 41(1):118–127

Cvijanovic G, Dozet G, Djukic V, Djordjevic S, Puzic G (2012) Microbial activity of soil during the inoculation of soya bean with symbiotic and free-living nitrogen fixing bacteria. Afr J Biotech 11(3):590–597

Cvijanovic G, Dozet G, Milosević N, Lalevic B (2010) The importance of micro-organisms in organic production. In: International tourism fair, Mediterranean days, Thematic proceedings from international conference "Healthy life", 1–8, Trebinje, Bosnia and Herzegovina, 08–10 April 2010, pp 86–93

Dhanasekar R, Dhandapani R (2012) Effect of biofertilizer on the growth of *Helianthus annuus*. Int J Plant Anim Environ Sci 2(4):134–137

Dianda M, Edeh E, Ukem B, Bala A, van Heerwaarden J, Giller KE (2016) Field crops research understanding variability in soybean yield and response to p-fertilizer and rhizobium inoculants on farmers' fields in Northern Nigeria. Field Crops Res 186:133–145

Doltra J, Lagdsmand M, Olesen JE (2011) Cereal yield and quality as affected by nitrogen availability in organic and conventional arable crop rotations: a combined modeling and experimental approach. Eur J Agron 34:83–95

Eghball B (2002) Soil properties as influenced by phosphorus- and nitrogen-based manure and compost applications. Agron J 94:128–135

El Sabagh A, Abd Omar E, Saneoka H, Barutçular C (2015) Evaluation of agronomic traits of canola (Brassica napus L.) under organic, bio- and chemical fertilizers. DUFED 4(2):59–67

El Sheikha AF (2016) Mixing manure with chemical fertilizers, why? and what is after? Nutr Food Technol 2(1):112. https://doi.org/10.16966/2470-6086

Elliott J, Deryny D, Muller C, Frieler K, Konzmann M, Gerten D, Glotter M, Florke M, Wada Y, Best N, Eisner S, Fekete BM, Folberth C, Foster I, Gosling SN, Haddeland I, Khabarov N, Ludwing F, Masaki Y, Olin S, Rosenzweig C, Ruane AC, Satoh Y, Schmid E, Stacke T, Tang Q, Wisser D (2013) Constraints and potentials of future irrigation water availability on agricultural production under climate change. Proc Natl Acad Sci 11:3239–3244

Endris S, Dawid J (2015) Yield response of maize to integrated soil fertility management on acidic nitosol of southwestern Ethiopia. J Agron 14:152–157

Ezung NK, Jami J (2019) Study on growth, yield and economics in maize as influenced by different levels of nitrogen applied through vermicompost and spacing in hill areas of North East Indian. J Soils Crops 29(2):236–242

Hafeez F, Hameed S, Zaidi AH, Malik KA (2002) Biofertilizers for sustainable agriculture. In: Azam F, Iqbal MM, Inayatullah C, Malik KA (eds) Techniques for sustainable agriculture. NIAB, Faisalabad, Pakistan, p 6773

Hai-cheng XU, Xing-long DAI, Jin-peng CHU (2018) Integrated management strategy for improving the grain yield and nitrogen-use efficiency of winter wheat. J Integr Agric 17:315–327

Hamidi M, Nasri M, Hanifepoor M (2015) Comparison of chemical and biological fertilizers on yield and yield components of RGS canola seed in different seed densities. Int J Agric Crop Sci 8(3):427–432

Hammed BT, Oloruntoba OE, Ana GREE (2019) Enhancing growth and yield of crops with nutrient-enriched organic fertilizer at wet and dry seasons in ensuring climate-smart agriculture. Int J Recycl Org Waste Agric 8(Suppl 1):S81–S92

Hartman LG, Pawlowski LM, Herman KT, Eastburn D (2016) Organically grown soybean production in the USA: constraints and management of pathogens and insect pests. Agronomy 6:16. https://doi.org/10.3390/agronomy6010016

Hayat R, Ali S, Amara U, Khalid R, Iftikhar A (2010) Soil beneficial bacteria and their role in plant growth promotion: a review. Ann Microbiol 60:579. https://doi.org/10.1007/s13213-010-0117-1

Hussain F, Malik AU, Haji MA, Malghani AL (2011) Growth and yield response of two cultivars of Mung bean (Vigna radiata L.) to different potassium levels. J Anim Plant Sci 21:622–625

Jenkinson DS (1988) The determination of microbial biomass carbon and nitrogen in soil. Advances in nitrogen cycling in agricultural ecosystems. CAB International, Wallingford, pp 368–386

Kazemeini SA, Hamzehzarghani H, Edalat M (2010) The impact of nitrogen and organic matter on winter canola seed yield and yield components. Aust J Crop Sci 4:335–342

Khaim S, Chowdhury MAH, Department BKS (2013) Organic and inorganic fertilization on the yield and quality of soybean. J Bangladesh Agril Univ 11(1):23–28

Khamparia NK, Thakur R, Sawarkar SD (2018) Effect of continuous use of inorganic fertilizers and organic manure on crop productivity, soil fertility and sustainability of soybean-wheat cropping system in a vertisol. J Soils Crops 28(1):19–25

Khan AA, Inamullah JMT (2014) Impact of various nitrogen and potassium levels and application methods on grain yield and yield attributes of wheat. Sarhad J Agric 30(1):35–46

Khan MN, Mobin M, Abbas ZK, Alamri SA (2018) Fertilizers and their contaminates in soils, surface and groundwater. In: Dominick A, Sala D, Goldstein MI (eds) The encyclopedia of the anthropocne, vol 5. Elsevier, Oxford, pp 225–240

Khandekar SD, Ghotmukale AK, Dambale AS, Suryawansh SB (2018) Growth, yield and economics of sunflower (helianthus annuusl.) as influenced by biofertilizer and fertilizers levels. Int J Curr Microbiol App Sci 6:1564–1571

Kombiok JM, Buah SSJ, Segbedji JM (2008). Enhancing soil fertility for cereal crop production through biological practices and the integration of organic and in-organic fertilizers in Northern Savanna Zone of Ghana. In: Soil fertility. INTECH Publication. https://doi.org/10.5772/53414

Korkmaz K (2007) Tar Õm Girdi Siste minde Azotve Azot Kirlilioi. http://www.ziraat.ktu.edu.tr/tarim_girdi.htm

Lal R (2011) Sequestering carbon in soils of agro-ecosystems. Food Policy 36:S33–S39

Maiorana M, Charfeddine M, Montemurro F, Vonella AV (2005) Reduction of agronomic inputs in sunflower (Helianthus annuus L.). HELIA 28(42):133–146

Mohamed SM, Mohamed HM, Shahata HM, Ahmed HM (2017) Growth, yield and yield components of canola crop (Brassica napus) in El-Kharga Oasis New Valley as affected by bio, nitrogen and phosphorus fertilization. Assiut J Agric Sci 48(1–1):319–330

Mohammadi K, Rokhzadi A (2012) An integrated fertilization system of canola (Brassica napus L.) production under different crop rotations. Indust Crop Prod 37:264–269

Mondini C, Sequi P (2008) Implication of soil C sequestration on sustainable agriculture and environment. Waste Manag 28(4):678–684

Moral R, Moreno-Caselles J, Perez-Murcia MD, Perez-Espinosa A, Rufete B, Paredes C (2005) Characterization of the organic matter pool in manures. Bioresour Technol 96(2):153–158

Namvar A, Khandan T, Shojaei M (2012) Effects of bio and chemical nitrogen fertilizer on grain and oil yield of sunflower (helianthus annuus l.) under different rates of plant density. Ann Biol Res 3(2):1125–1131

Oldroyd GED, Dixon R (2014) Biotechnological solutions to the nitrogen problem. Curr Opin Biotechnol 26:19–24

Palm CA, Gachengo CN, Delve RJ et al (2001) Organic inputs for soil fertility management in tropical agroecosystems: application of an organic resource database. Agric Ecosyst Environ 83:27–42

Patidar M, Malhi AI (2004) Effect of farmyard manure fertility levels and biofertilizers on growth, yield and yield quality of sorghum. Indian J Agron 49:117–120

Posner JL, Baldock JO, Hedtcke JL (2008) Organic and conventional production systems in Wisconsin integrated cropping systems trials. I. Productivity 1990–2002. Agron J 100:253–260

Rahman L (1982) Cultivation of soybean and its uses. City Press, Dhaka, pp 5–7

Raj MA, Vasanthi D (2019) Effect of organic amendments and inorganic fertilizers on the properties of theri soil. J Soils Crops 29(1):78–82

Rivero C, Tchirenje L, Ma Q, Martinez G (2004) Influence of compost on soil organic matter quality under tropical conditions. Geoderma 123:355–361

Serafim B, Oginga B, Mugwe JN (2013) Effects of manure, lime and mineral P fertilizer on soybean yields and soil fertility in a *humic nitisol* in the central highlands of Kenya. Int J Agric Sci Res 2:283–291

Sharaan AN, Gallab KH (2002) Selection in canola (Brascianapus L.) germplasm under conditions of newly reclaimed land. 1: Variability and genetic parameters in the base lines Egypt. J Plant Breed 6(2):1–13

Simpson RJ, Oberson A, Culvenor RA, Ryan MH, Veneklaas EJ, Lambers H, Lynch JP, Ryan PR, Delhaize E, Smith FA, Smith SE, Harvey PR, Richardson AE (2011) Strategies and agronomic interventions to improve the phosphorus-use efficiency of farming systems. Plant Soil 349:89–120. https://doi.org/10.1007/s11104-011-0880-1

Sutton MA, van Grinsven H, Billen G, Bleeker A, Bouwman AF, Bull K, Erisman JW, Grennfelt P, Grizzetti B, Howard CM, Oenema O, Spranger T, Winiwarter W (2011) Summary for policy makers. In: Sutton MA, Howard CM, Erisman JW, Billen G, Bleeker A, Grennfelt P, van Grinsven

H, Grizzetti B (eds) The European nitrogen assessment. Cambridge University Press, pp xxiv–xxxiv

Tomm GO, Ferreira PEP, Aguiar JLP, Castro AMG, Lima SMV, Mori C (2010) Panorama atual e indicações para aumento de eficiência da produção de canola no Brasil. EmbrapaTrigo, Passofundo, p 82

Vanlauwe B, Bationo A, Chianu J, Giller KE, Merckx R, Mokwunye U, Ohiokpehai O, Pypers P, Tabo R, Shepherd K, Smaling E, Woomer PL, Sanginga N (2010) Integrated soil fertility management: operational definition and consequences for implementation and dissemination. Outl Agric 39:17–24

Usman M, Madu VU, Alkali G (2015) The combined use of organic and inorganic fertilizers for improving maize crop productivity in Nigeria. Int J Sci Res Pub 5:1–7

Wakeel A, Farooq M, Qadir M, Schubert S (2011) Potassium substitution by sodium in plants. Crit Rev Plant Sci 30:401–413. https://doi.org/10.1080/07352689.2011.587728

Woźniak A, Soroka M (2018) Effect of crop rotation and tillage system on the weed infestation and yield of spring wheat and on soil properties. Appl Ecol Environ Res 16:3087–3096

Wu W, Ma B (2015) Integrated nutrient management (INM) for sustaining crop productivity and reducing environmental impact a review. Sci Total Environ 512–513:415–427

Zerihun A, Abera T, Dedefo T, Fred K (2013) Maize yield response to crop rotation, farmyard manure and inorganic fertilizer application in Western Ethiopia. Afr J Agric Res 8:5889–5895

Zingore S, Mutegi J, Agesa B et al (2015) Soil degradation in sub-Saharan Africa and crop production options for soil rehabilitation. Better Crops Plant Food 99:65–67

Zohry A, Mohamadin AA (2020) Chemical versus nonchemical fertilizer sources to enhance the yield of three edible oil crops in East El-Owinate region in Egypt. J Soils Crops 30(2):191–199

Chapter 3
Fish Farms Effluents for Irrigation and Fertilizer: Field and Modeling Studies

Abd El-Hafeez Zohry and Samiha Ouda

Abstract This chapter presented an integration of water-smart and soil-smart practices to sustainably increase crops production. The first part of the chapter discussed the benefits of partial or full replacement of irrigation water and fertilizer with fish farms effluents applied to winter and summer crops interplanted between orange trees in four crop sequences. This part was implemented in a field experiments. The second part of the chapter used modeling to determine the possibility of using the effluents of several fish farms exerted from one area in Egypt to expend on cultivating new areas and increase food availability. This chapter could provide a work plan to sustain soil and water resources through the cultivation of different crop sequences and irrigation with fish farms effluents. Doing that could improve soil quality, reduce environmental pollution by fish farms wastes, reduce fertilizer usage and groundwater contamination, reduce weeds spread and usage of pesticides, thus reducing production costs.

Keywords Faba bean · Lupine · Soybean · Sunflower · Water and land equivalent ratios

1 Introduction

Water resource pollution has been identified as a hazardous problem affecting mankind around the world, with many health and economic consequences. It has a significant risk not only on human health, but also on animal health and agricultural production (Dakkak 2016). Agricultural production quality and quantity has been affected by using untreated water for irrigation. Agricultural runoffs containing fertilizers, pesticides, industrial effluents and untreated sewage (El Bedawy 2014).

A. E.-H. Zohry (✉)
Crops Intesefication Research Department, Field Crops Research Institute; Agricultural Research Center, Giza, Egypt

S. Ouda
Water Requirements and Field Irrigation Research Department, Soils, Water and Environment Research Institute; Agricultural Research Center, Giza, Egypt

S. Ouda and A. E.-H. Zohry, *Climate-Smart Agriculture*,
https://doi.org/10.1007/978-3-030-93111-7_3

Although developing countries are characterized by a lower industrialization, the deficiency observed in solid waste and wastewater management and the lack of city planning constitute factors facilitating the environmental pollution of cities (Maillard and Santos 2008). Moreover, unexpected population growth is related to water quality degradation and is causing large increase in nutrients and microbial loads (Zhang et al. 2009).

As water deficiency in Egypt is getting worst, it is crucial to define actions to face it. In some areas where fish farms exist in Egypt, some farms are dumping its effluents in the water resources. To prevent that and increase food production it could be used in irrigation after testing it for microbial diseases. Fish farms effluent is a neglected water resource in Egypt, if wisely used, it can contribute to increasing food production. Irrigating crops with this type of water could relieve pressure on freshwater resources (Fedoroff et al. 2010), and improving soil quality due to organic matter that exists in fish water (Elnwishy et al. 2006), and that could help in reducing the contamination of the environment.

Moreover, it could be a source of irrigation water and organic materials to be used in low fertile soils. Isitekhale and Adamu (2016) reported that only 25% of N and 20% of P of fish feed is recovered in harvested fish and the rest is accumulated in farm effluent. Incorporating these organic wastes into the low fertile soil helps to build up soil organic matter layers needed for steady supply of nutrients to the growing plants (Zieman et al. 2007), increased the yield of cultivated crops and improve soil chemical properties, such as available phosphorus, water soluble potassium, calcium and magnesium (Ojobor and Tobih 2015). Better plant growth and higher productivity were reported under irrigation with fish farms effluent as a result of improvement in roots uptake of dissolved nutrients excreted directly by fish or produced from the microbial breakdown of fish wastes (Elnwishy et al. 2006). Positive effects of using fish farm effluent in irrigating potato, soybean, and onion were observed on their yield (Haquea et al. 2016). Castro et al. (2006) reported that the yield of tomato irrigated with fish water effluent increased in the first three harvestings.

This chapter presented the effect of the integration of water-smart and soil-smart practices in increasing food availability. The first part of the chapter discussed the benefits of partial or full replacement of irrigation water and fertilizer with fish farms effluent applied to winter and summer crops interplanted between orange trees in four crop sequences, with respect to yield of each crop and orange. This part was implemented in a field experiments. The second part of the chapter used modeling to determine the possibility of using the effluents of several fish farms exerted from one area in Egypt to expend on cultivating new areas and increase food availability.

2 Studies on Using Fish Farms Effluents for Irrigation in Egypt

The total area of fish farms in Egypt is around 179.2 thousand hectares spreads over 19 locations in Egypt; produces 705.5 thousand tons of fish every year. Field experiments were conducted in Egypt to study the benefits of using fish farms effluents in irrigation. Abdelraouf and Hoballah (2014) reported that soybean yield irrigated with 100% of fish water effluent and 100% nitrogen fertilizer was higher by 50% than soybean yield irrigated with 100% Nile water and 100% nitrogen fertilizer. Abdelraouf and Ragab (2017) indicated that wheat yield was increased by 10% under application of 100% fish water effluent and 100% nitrogen fertilizer, compared to application of 100% Nile water with 100% nitrogen fertilizer. Another study was conducted in Wadi El-Natron, Behira Governorate in North Egypt by AbdEl Magid et al. (2018), where they compared between the productivity of onion and sugar beet irrigated with farm effluents and their productivity when irrigated with Nile water. They reported that using fish farm effluent in irrigating these two crops increased their yields by 16% without adding fertilizers. Another study was done by a group of scientists from WorldFish Center located in Abbass; Sharkia governorate; Egypt in three governorates provided information on fish production, water use, and cultivated crops (van der Heijden 2012). The study concluded that these farms used water most efficiently in fish production, as well as in fruits and crops irrigation.

Zohry et al. (2020a) indicated that the yield of faba bean and chickpea irrigated with fish farm effluent increased by 11 and 29%, respectively, compared to irrigation with Nile water. Besides, interplanting faba bean, chickpea, lupine, and lentil under orange trees and irrigation with fish farm effluents increased orange yield by 14, 14, 18, and 15%, respectively, compared to irrigation with Nile water (Zohry et al. 2020a).

3 Benefits of Irrigation with Fish Farms Effluents and Implementing Intercropping Systems

To increase water productivity of using fish farm effluents, one procedure could be done, namely choosing the appropriate intercropping system. Implementing intercropping systems proved to enhance land and water productivity (Ouda et al. 2018; Zohry and Ouda 2019). Said et al. (2016) indicated that implementing intercropping systems, where two crops sharing the area and the applied water to one of them is an alternative to cope with water security. It was reported that water productivity, water use efficiency, and water equivalent ratio were increased under implementing intercropping systems (Zohry et al. 2017; El-Mehy et al. 2018; Ouda et al. 2018; Zohry and Ouda 2019). Harvested yield of two crops under intercropping systems is the main factor in increasing land productivity (Zohry et al. 2017), in addition to increasing soil fertility by intercropping with legumes (Zohry et al. 2020b), minimize predator's infestation (Hamd-Alla et al. 2014), and improving the microclimate of the intercropping system for the benefits of the final productivity (Kestha and El-Baz 2004).

4 Integration of Water-Smart and Soil-Smart Practices

I. Field study

Interplanting between fruit trees is a recommended procedure to use the spaces between fruit trees in growing economic crops. The interplanted crops could act as a cover crop to prevent weeds growing and increasing water and land productivities (Bakshi et al. 2019). Abdel-Aziz et al. (2008) reported that orange fruit yield was enhanced and fruit drop was decreased under interplanting with legume crops. El-Mehy and El-Badawy (2017) indicated that interplanting soybean between orange trees can substitute for 25% of the recommended NPK, resulted in an increase in the yield of both crops. Selim et al. (2020a) interplanted soybean in two plant distributions (wide and narrow distribution) with the same planting density between orange trees. They found that wide soybean plant distribution increased orange fruit yield per hectare by 10%, in addition to increase land equivalent ratio and total return, compared to solid planting of orange trees. Thus, there is a potential benefits from interplanting between orange trees in Egypt.

A field experiment was conducted in El-Kassaseen Agricultural Experiments Research Station (Lat. 30° 35′ 30″ N, Long. 32° 14′ 50″ E, 10 m a.s.l.), Agricultural Research Center, Ismailia Governorate, Egypt during 2016/17 and 2017/18 seasons for field crops and 2018 and 2019 for orange. The aim of the study was to assess the effect of three irrigation treatments on interplanting faba bean or lupine between 3-year old orange trees in two separate experiments using a strip plot design. These two winter interplanting systems were followed by interplanting soybean or sunflower between the orange trees in the summer season in two separate experiments using also a strip-split plot design taking into account the effect of preceding winter crops. The effect of the four crop sequences on orange yield was studied after its harvest in 2018 and 2019 (Fig. 1). The soil of the experiment was sandy, with pH value = 7.6 and EC value = 0.56 dS/m. Maximum temperature in the winter was between 17–22 °C and in the summer was between 25–35 °C.

5 Applied Irrigation Water to the Studied Interplanting Systems

The effect of three irrigation treatments Irrigation I (full irrigation using Nile water and recommended NPK fertilizer (control); Irrigation II (full irrigation using 50% fish farm effluent and 50% Nile water, in addition to 50% of recommended fertilizer)

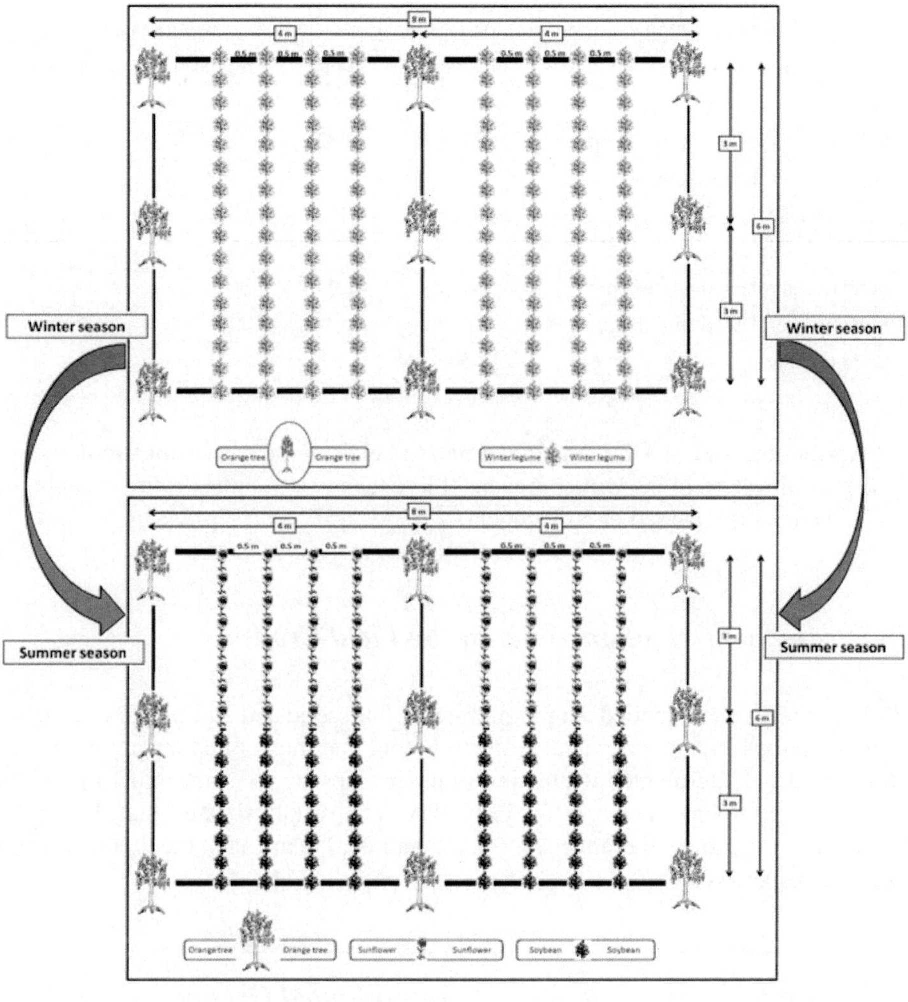

Fig. 1 Experimental plots of winter crops interplanted between orange trees (**a**) followed by summer crops interplanted between orange trees (**b**)

and Irrigation III (full irrigation using fish farm effluent without fertilizer) were tested on interplanting winter crops (faba bean and lupine) followed by summer crops (soybean and sunflower) between orange trees in four crop sequences.

The required irrigation amount for winter crops (solid or interplanted) was 100% of ETo and for summer crops (solid or interplanted) was 120% ETo. Solid orange was irrigated with 120% ETo. With respect to interplanted orange, 70% of its required irrigation amount was applied during the growth cycle of interplanted winter and summer crops. It was assumed that the lateral movement of irrigation water from the interplanted crops will compensate the remaining 30% of orange water requirements. In addition, full irrigation was applied in the period between the harvest of winter crops and the cultivation of summer crops (more than one month, from the beginning

Table 1 Applied irrigation water (m^3/ha) to the studied crops in both growing seasons

Applied irrigation water	Faba bean	Lupine	Soybean	Sunflower
First season				
Solid or interplanted winter crops	4,788	4,282	–	–
Solid or interplanted summer crops	–	–	7,765	7,529
Solid Orange	20,965	–	–	–
Second season				
Solid or interplanted winter crops	4,741	4,235	–	–
Solid or interplanted summer crops	–	–	7,894	7,635
Solid orange	21,059	–	–	–

of April to the mid of May) and in the period between the harvest of summer crops and the cultivation of the winter seasons (more than two months, from the beginning of September to the med of November).

5.1 Applied Irrigation Water to the Field Crops

Table 1 presents the applied irrigation water to both solid and interplanted winter and summer crops and solid orange. It worth noting that the applied irrigation water to both solid and interplanted winter and summer crops are the same under the studied irrigation treatments, but it differ in quality. The results showed that the applied irrigation water to solid orange was 20,965 and 21,059 m^3/ha in the first and second season, respectively.

5.2 Applied Irrigation Water to Interplanted Orange

Table 2 showed that, under interplanting conditions of orange, the applied irrigation water was divided into four periods. In the first and the third periods, either a winter or a summer crop was interplanted between orange trees, orange trees received 70%

Table 2 Applied irrigation water (m^3/ha) to interplanted orange in the first and second season

	Interplanted orange	First season	Second season
Period 1	Interplanted winter crops under orange trees	4,859	4,871
Period 2	In between winter and summer crops	2,506	2,624
Period 3	Interplanted summer crops under orange trees	4,424	4,447
Period 4	In between summer and winter crops	3,671	3,800
Total		15,459	15,741

of its required water. In the second and the fourth periods, orange had its complete water requirements, where there were no field crops interplanted. These periods were in between the harvest of a winter crop until the cultivation of the following summer crop, and after the harvest of a summer crop until the cultivation of the following winter crop. Therefore, the amount of saved irrigation water to orange under intercropping conditions were 5,600 m^3/ha (the difference between its value in Table 1 and the value in Table 2 in the first season) and 5,318 m^3/ha (the difference between its value in Table 1 and the value in Table 2 in the second season) or 26 and 25%, respectively of its required amount under solid cultivation in the first and second season, respectively (Table 2). These saved irrigation water could contribute in irrigating other areas and increase food production.

5.3 Applied Irrigation Water to the Studied Crop Sequences

The results in Table 3 showed that the highest amount of irrigation water was applied to faba bean followed by soybean both interplanted between orange trees. Whereas, the lowest amount of irrigation water was applied to lupine followed by sunflower both interplanted between orange trees. The table also showed that there was low differences in the applied irrigation water to each crop sequence in each season, which encouraged the cultivation of either of them.

Table 3 Total applied irrigation water (m^3/ha) to each crop sequence in the first and second season

Crop sequence	Total applied irrigation water
First season	
Interplanted faba bean + interplanted soybean between orange trees	28,012
Interplanted faba bean + interplanted sunflower between orange trees	27,776
Interplanted lupine + interplanted soybean between orange trees	27,506
Interplanted lupine + interplanted sunflower between orange trees	27,271
Second season	
Interplanted faba bean + interplanted soybean between orange trees	28,376
Interplanted faba bean + interplanted sunflower between orange trees	28,118
Interplanted lupine + interplanted soybean between orange trees	27,871
Interplanted lupine + interplanted sunflower between orange trees	27,612

6 The Yield of the Studied Field Crops

6.1 Winter Crops

The results in Table 4 showed that both faba bean and lupine yield was higher in the second season, compared to the first season under the three irrigation treatments. This could be a result of decomposition of the roots of both crops after its harvest in the first season, which could increase organic matter in the rhizosphere as indicated by Li et al. (2003). They also stated that intercropped legumes tends to fix more atmospheric N_2 than monoculture; as a result of competition between the interplanted crops in the use of N_2 and that has positive effects on the organic matter in the rhizosphere.

Furthermore, faba bean yield was increased by 4 and 6% for interplanted and solid faba bean in the first season under Irrigation II and III, respectively, compared to the values obtained under Irrigation I. Whereas, in the second season, the increase was 9 and 0% for the interplanted faba bean under the same respective treatments, compared to the values obtained under Irrigation I (Table 4). This increase in the yield of faba bean could be attributed to the high amount of N and P in the fish farm effluents, as stated by Isitekhale and Adamu (2016).

With respect to lupine, Table 4 showed that interplanted lupine yield was increased by 2 and 4% under Irrigation II and III in the first season, respectively and in the second season it was increased by 5 and 12%, under the same treatments, compared to its value under Irrigation I.

Table 4 The yield of the studied winter crops as affected by the interaction between irrigation treatments and cropping systems

Irrigation treatment	Cropping system	Faba bean yield (ton/ha)		Lupine yield (ton/ha)	
		First season	Second season	First season	Second season
Irrigation I	Intercrop	1.12	1.14	2.85	2.79
	Solid	2.90	2.98	6.79	7.09
	Mean	2.01	2.06	4.82	4.94
Irrigation II	Intercrop	1.20	1.29	2.94	2.97
	Solid	3.03	3.19	6.91	7.45
	Mean	2.11	2.25	4.92	5.21
Irrigation III	Intercrop	1.22	1.31	2.98	3.25
	Solid	3.08	3.22	7.01	7.80
	Mean	2.15	2.06	4.99	5.52

Irrigation I: Full irrigation using Nile water and 100% NPK fertilizer (control), Irrigation II: Full irrigation using 50% fish farm effluent and 50% Nile water, in addition to 50% NPK fertilizer, Irrigation III: Full irrigation using fish farm effluent

6.2 Summer Crops

6.2.1 Soybean

The results in Table 5 indicated that, among the two preceding winter crops (faba bean or lupine), the highest interplanted soybean yield was obtained when lupine preceded it under Irrigation II and III, compared to Irrigation I. Soybean yield was increased by 3 and 5%, in the first season and by 5 and 7% in the second season under Irrigation II and III, respectively compared to its value under Irrigation I. Hauggaard-Nielsen (2009) indicated that the residual N_2 left in the soil by lupine N_2 fixation is higher than its counterpart value fixed by faba bean. Furthermore, lupine has highly efficient P-acquisition capacity (Weisskopf et al. 2011), which is important for nodulation and N_2 fixation (Steingrobe 2005). This high efficient P-acquisition capacity is related to lupine roots characteristics, which allow to exude P-solubilizing compounds, namely citrate that release P from mineral compounds and acid phosphatase that release P from organic compounds (Gilbert et al. 1999).

The results also showed that the highest soybean yield was obtained under Irrigation III in both growing seasons. It seem that the presence of nitrogen fixing bacteria from soybean may facilitate the availability of nutrients from the organic material existed in the fish farm effluent and these nutrients were enough for soybean to produce the highest yield. Haquea et al. (2016) observed positive effects of using fish water effluent for irrigation on soybean yields, compare to irrigation with conventional water.

Table 5 Soybean yield as affected by the interaction between irrigation treatments, preceding winter crops and cropping systems

Water qualities	Preceding crop	First season			Second season		
		Inter	Solid	Mean	Inter	Solid	Mean
Irrigation I	Faba bean	1.35	3.28	2.31	1.38	3.35	2.36
	Lupine	1.39	3.35	2.37	1.45	3.39	2.42
	Mean	1.37	3.31	2.34	1.41	3.37	2.39
Irrigation II	Faba bean	1.40	3.33	2.36	1.50	3.48	2.49
	Lupine	1.48	3.41	2.44	1.57	3.48	2.52
	Mean	1.44	3.37	2.4	1.53	3.48	2.5
Irrigation III	Faba bean	1.45	3.40	2.42	1.53	3.52	2.52
	Lupine	1.50	3.47	2.48	1.58	3.57	2.57
	Mean	1.47	3.43	2.45	1.55	3.54	2.55

Irrigation I: 100% Nile water + 100% NPK fertilizer (control), Irrigation II: 50% fish farm effluent + 50% Nile water + 50% NPK fertilizer and Irrigation III: 100% fish farm effluent only

6.2.2 Sunflower

The highest sunflower yield was obtained when lupine preceded it, compared to faba bean preceding it in both seasons and under Irrigation II, compared to Irrigation I and III Table 6. This result implied that the application of Irrigation III did not fulfill sunflower's needs of nutrients, due to unavailability of nitrogen fixing bacteria, as in soybean, to facilitate nutrients accessibility. Similar results were obtained by Selim and Shams (2019) for maize interplanted between mandarin trees and irrigated with fish farm effluent.

Table 6 also showed revealed that there was low difference between interplanted and solid sunflower yield, although the planting density of the interplanted sunflower was 50% of its solid planting density. This could be attributed to nitrogen residual after the preceding legume crop (faba bean or lupine) contributed in an increase in sunflower yield as reported by Hauggaard-Nielsen et al. (2009). In addition, interplanting sunflower plants between orange trees is protected from the winds, which allows more visitations of bees; the most important insects in sunflower pollination process (Morgado et al. 2002).

6.3 *Orange Yield*

As it was stated earlier, a winter crop and a summer crop were interplanted between orange trees, therefore the obtained orange yield is affected by their sequence. The results in Table 7 revealed that the crop sequence that attained the highest orange fruit yield per hectare was lupine followed by soybean both interplanted between orange

Table 6 Sunflower yield as affected by the interaction between irrigation treatments, preceding winter crops and cropping systems

Irrigation treatments	Preceding crop	First season			Second season		
		Interplanted	Solid	Mean	Interplanted	Solid	Mean
Irrigation I	Faba bean	1.00	1.08	1.04	1.12	1.19	1.15
	Lupine	1.05	1.11	1.08	1.18	1.21	1.19
	Mean	1.02	1.09	1.06	1.15	1.20	1.17
Irrigation II	Faba bean	1.06	1.09	1.07	1.16	1.20	1.18
	Lupine	1.14	1.18	1.16	1.26	1.30	1.23
	Mean	1.08	1.11	1.09	1.19	1.22	1.20
Irrigation III	Faba bean	1.07	1.11	1.09	1.18	1.22	1.20
	Lupine	1.12	1.16	1.14	1.22	1.25	1.28
	Mean	1.10	1.14	1.12	1.22	1.26	1.24

Irrigation I: 100% Nile water + 100% NPK fertilizer (control), Irrigation II: 50% fish farm effluent + 50% Nile water + 50% NPK fertilizer and Irrigation III: 100% fish farm effluent only

Table 7 Effect of the interaction between irrigation treatments, preceding winter crops, and cropping systems on orange yield in 2018 and 2019 seasons

Treatments	Preceding crops	Orange yield in the first season			Orange yield in the second season		
		Soybean	Sunflower	Mean	Soybean	Sunflower	Mean
Irrigation I	Faba bean	2.80	2.60	2.71	2.83	2.65	2.77
	Lupine	2.90	2.60	2.81	2.94	2.68	2.70
	Mean	2.85	2.60	2.76	2.89	2.67	2.74
Irrigation II	Faba bean	2.90	2.68	2.79	3.00	2.75	2.87
	Lupine	2.95	2.71	2.86	3.01	2.71	2.83
	Mean	2.88	2.64	2.79	2.93	2.69	2.78
Irrigation III	Faba bean	2.96	2.76	2.86	3.34	2.80	3.07
	Lupine	3.38	2.76	3.07	3.60	2.93	3.27
	Mean	3.17	2.76	2.96	3.47	2.86	3.17

Irrigation I: 100% Nile water + 100% NPK fertilizer (control), Irrigation II: 50% fish farm effluent + 50% Nile water + 50% NPK fertilizer and Irrigation III: 100% fish farm effluent only

trees irrigated with Irrigation III in both season, where orange yield was increased by 17 and 25%, in the first and second season, respectively, compared to the value obtained under irrigation I. This result could be attributed to the root characteristics of lupine and soybean (Watt and Evans 2003), which allowed to the release of more P from organic compounds in the soil (Gilbert et al. 1999). Thus, cultivation of lupine followed by soybean might increase the amount of P in the soil, as well as facilitate its absorption by the roots of orange trees and that resulted in the increase in orange yield under that crop sequence, compared to the other studied sequences. Direcenmattos et al. (2010) indicated that the availability of P in the soil resulted in greater growth and yield of citrus trees corresponded to greater root development and growth rate.

Furthermore, it can be noticed from the table that the orange yield was higher in the second growing season, compared to its value in the first season under the three irrigation treatments.

7 Water and Land Equivalent Ratios of the Studied Crop Sequences

7.1 *Water Equivalent Ratio (WER)*

WER was used to quantify the efficiency of water use by each crop sequence interplanted between orange trees. It is defined as the total water needed in sole crops to produce the equivalent amount of crop yield on a unit area of intercrop (Mao et al. 2012). WER was calculated for each crop sequence interplanted under orange trees as follows:

$$WER = \frac{\left(\frac{Y_{int,w}}{WU_{int,w}}\right)}{\left(\frac{Y_{mono,w}}{WU_{mono,w}}\right)} + \frac{\left(\frac{Y_{int,s}}{WU_{int,s}}\right)}{\left(\frac{Y_{mono,s}}{WU_{mono,s}}\right)} + \frac{\left(\frac{Y_{int,O}}{WU_{int,O}}\right)}{\left(\frac{Y_{mono,O}}{WU_{mono,O}}\right)} \quad (1)$$

where $Y_{int,w}$, $Y_{int,s}$ and $Y_{int,O}$ are the yield of interplanted winter crop, summer crop and orange, respectively. $WU_{int,w}$, $WU_{int,s}$ and $WU_{int,O}$ are water consumptive use by the interplanted winter crop, interplanted summer crop and interplanted orange. $Y_{mono,w}$, $Y_{mono,s}$ and $Y_{mono,O}$ are the yield of mono winter crop, mono summer crop and mono orange, respectively. $WU_{mono,w}$, $WU_{mono,s}$ and $WU_{mono,O}$ are water consumptive use by mono winter crop, mono summer crop and mono orange, respectively.

If the WER > 1, it suggests that the water utilization of interplanting is higher than that of monoculture and that imply advantage in implemented interplanting system. If WER < 1, it shows that water utilization of interplanting is lower than that of monoculture and that imply disadvantage.

The results in Table 8 indicated that the values of WER for orange interplanting systems with either winter or summer crops were higher than 1.0 under the three irrigation treatments in both growing seasons. The highest value of total WER was found for crop sequence: lupine followed by sunflower both interplanted between orange trees irrigated with Irrigation III in first and second seasons, respectively namely 2.60 and 2.59. This result is attributed to the lowest required irrigation water for that crop sequence as shown in Table 3. The results also implied that the water utilization of this crop sequence were higher than the value of the other crop sequences. In this case, the values of WER_{total} were increased by 160 and 159% for that crop sequence irrigated with Irrigation III in the first and second growing seasons, respectively.

7.2 Land Equivalent Ratio (LER)

LER is the ratio of the area under sole cropping to the area under interplanting needed to give an equal amounts of yield at the same management level. It is the sum of the fractions of the interplanted yields divided by the sole crop yields (Mead and Willey 1980). LER is calculated for each crop sequence interplanted between orange trees as follows:

$$LER = (Y_{WI}/Y_{WW}) + (Y_{SI}/Y_{SS}) + (Y_{OI}/Y_{OO}) \quad (2)$$

where Y_{WI}, Y_{SI} and Y_{OI} = Interplanted yield of interplanted winter crop, summer crop and orange, Y_{WW}, Y_{SS} and Y_{OO} = Pure stand yield of winter crop, summer crop and orange.

If the LER > 1, it suggests that the land utilization of interplanting is higher than that of monoculture. If LER < 1, it shows that land utilization of interplanting is lower than that of monoculture.

Table 8 Water equivalent ratio (WER) for the studied crop sequences

Irrigation treatments	Preceding crops	Crop WER				Total WER	
		Legume	Soybean	Sunflower	Orange	$WER_{Legume + soybean + orange}$	$WER_{Legume + sunflower + orange}$
First season							
Irrigation I	Faba bean	0.39	0.48	0.93	1.18	2.04	2.49
	Lupine	0.42	0.41	0.95	1.18	2.01	2.54
Irrigation II	Faba bean	0.40	0.49	0.93	1.18	2.07	2.51
	Lupine	0.43	0.43	0.95	1.20	2.06	2.57
Irrigation III	Faba bean	0.40	0.51	0.94	1.19	2.09	2.53
	Lupine	0.43	0.43	0.97	1.21	2.07	2.60
Second season							
Irrigation I	Faba bean	0.42	0.41	0.94	1.15	1.98	2.51
	Lupine	0.39	0.43	0.98	1.17	1.99	2.54
Irrigation II	Faba bean	0.44	0.43	0.97	1.15	2.03	2.56
	Lupine	0.40	0.45	0.98	1.18	2.03	2.56
Irrigation III	Faba bean	0.44	0.43	0.97	1.17	2.05	2.58
	Lupine	0.42	0.52	0.99	1.19	2.12	2.59

Irrigation I: 100% Nile water + 100% NPK fertilizer (control), Irrigation II: 50% fish farm effluent + 50% Nile water + 50% NPK fertilizer and Irrigation III: 100% fish farm effluent only

The results in Table 9 indicated that the values of LER for all the studied crop sequences were found to be greater than 1.0. The highest LER value was found for the following crop sequence: lupine followed by sunflower both interplanted under orange trees irrigated with Irrigation III in the first growing season, namely 2.61. Whereas, in the second growing season, the highest LER value was found for the following crop sequence: faba bean followed by sunflower both interplanted under orange trees irrigated with Irrigation III, namely 2.57.

These two crop sequences achieved yield advantages for orange and attained efficient utilization of land resources. The above result could be attributed to low differences between interplanted and solid sunflower yield (included in the calculation of LER), which resulted in higher LER. Similar results were reported by Selim and Shams (2019) and Selim et al. (2020a, b) for different interplanting systems.

II. Modeling irrigation with fish farms effluents in Wadi El-Natron in Egypt

Wadi El-Natron (30.43° N, 30.24° E) is located in El-Behira governorate in the west desert of Egypt. It is a depression in north Egypt located 23 m below sea level and 38 m below the Nile River level. The area has a potential for expansion for agricultural purposes (Gad and El-Zeiny 2016). In addition, there are many fish farms scattered there.

8 Estimation of Potential Irrigated Lands by Fish Farms Effluents

The published data by AbdEl Magid et al. (2018) reported that the area of fish farms (public and private farms) is 48.1 thousand hectares and uses 455.0 million cubic meters. These amounts of water are used in one fish production cycle. Each farm produces two production cycles of fish, 5 months each. The fish farms were designed to keep the water level fixed, where an amount of water is drained every day and replaced by a new amount. An average value of fish production per cycle averaged over three farms nearby the studied area was used in the calculation, namely 10.0 ton/hectare (van der Heijden 2012). Accordingly, the total fish production from the area of the studied fish farms can be calculated as followed:

$$Total\ fish\ production = Fish\ production * Area\ of\ fish\ farms\ (ton/ha) \quad (3)$$

The expected water loss as a result of the production of 1.0 tons of fish, in addition to evaporation from the water surface, is 29 m^3/cycle (personal communication). Total water losses during fish production in one fish cycle can be calculated as followed:

$$Total\ water\ losses = Total\ water\ in\ fish\ farms * Water\ losses * Cycle\ length\ \left(m^3/cycle\right) \quad (4)$$

Table 9 Land equivalent ratio for the studied crop sequences

Irrigation treatments	Preceding crops	Relative yield				LER	
		Legume	Soybean	Sunflower	Orange	LER $_{legume + soybean + orange}$	LER $_{legume + sunflower + orange}$
First season							
Irrigation I	Faba bean	0.38	0.42	0.97	1.03	1.84	2.39
	Lupine	0.42	0.43	0.96	1.11	1.97	2.50
Irrigation II	Faba bean	0.39	0.41	0.92	1.16	1.97	2.48
	Lupine	0.41	0.41	0.94	1.19	2.20	2.55
Irrigation III	Faba bean	0.39	0.42	0.96	1.18	2.00	2.54
	Lupine	0.42	0.43	0.96	1.22	2.08	2.61
Second season							
Irrigation I	Faba bean	0.38	0.43	0.96	1.12	1.93	2.47
	Lupine	0.39	0.45	0.97	1.09	1.94	2.47
Irrigation II	Faba bean	0.40	0.41	0.94	1.12	1.94	2.47
	Lupine	0.39	0.42	0.97	1.10	1.93	2.47
Irrigation III	Faba bean	0.41	0.43	0.96	1.19	2.04	2.57
	Lupine	0.41	0.44	0.96	1.16	2.02	2.55

Irrigation I: 100% Nile water + 100% NPK fertilizer (control), Irrigation II: 50% fish farm effluent + 50% Nile water + 50% NPK fertilizer and Irrigation III: 100% fish farm effluent only. Solid orange yield equal 2.29, 2.40, and 2.76 ton/ha in the first growing season and it were 2.31, 2.55 and 2.85 ton/ha in the second growing season

Table 10 Calculations of available amount of fish water effluent for crop production

Item	Amount
Total water of fish farms (m^3/cycle)	455,000,000
Area of fish farms (ha)	48,093
Fish production (ton/ha)	10.0
Total fish production (ton/cycle)	480,929
Amount of water losses (m^3/cycle)	2,092,042
The available amount of fish water (m^3/cycle)	452,907,958

The available water for crop production can be calculated by subtracting total water losses from total amounts of water used in fish farms as followed:

$$Available\,water\,for\,crop\,production = Total\,water\,in\,fish\,farms - Total\,water\,losses\left(m^3\right) \quad (5)$$

Table 10 presents the collected data used in the calculations of available water for crops production. The table indicated that water losses from the studied fish farms area could reach 2.1 million cubic meters. Thus, it allows having 452.9 million cubic meters per fish cycle available for irrigation. This amount could be invested in cultivated large areas nearby these fish farms.

9 The Suggested Cropping Systems

The potential of using the effluents of these farms in crops production was assessed. Three production alternatives were suggested and the expected cultivated area was estimated. Instead of mono-crops cultivation, we suggested implementing two crop rotations each have diversified crops. Additionally, the suggested third cropping system was the cultivation of suitable orchard fruit trees (evergreen or deciduous) as permanent crops and implementing interplanting of legume crops between it to increase land productivity.

9.1 Crop Rotations

The suggested crop rotations should be characterized by having intercropping systems, mostly with legumes to obtain diversified production that could increase food availability and reduce food gaps in Egypt, as well as improve soil fertility. The existed food gaps in Egypt are related to the shortage in the strategic crops, namely wheat, maize, faba bean (also other legumes), sugarcane (sugar beet is also considered), cotton, and rice. The suggested intercropping systems were previously studied and implemented in Egypt to assert their suitability and applicability under the environmental conditions of the area.

Furthermore, the suggested rotations contained crops tolerant to medium levels of salinity. Each rotation will be implemented in three years, with a total area of three hectares. In each year, winter and summer crops are included and rotating randomly in the following two years. These crop rotations will be irrigated with fish farms effluents. Zohry and Ouda (2018) have documented the advantages attained from the application of crop rotations on soil, crop production, and amounts of water used in irrigation, especially when intercropping systems are included. The suggested crop rotations contained most of these strategic crops in intercropping systems.

9.1.1 First Crop Rotation

The first suggested crop rotation was designed as legume crops rotation, mostly included in intercropping systems, where legume crops are dominated (Table 11). It included 36% legumes (total number of legumes divided by the total number of crops), 27% cereals, 18% sugar beet (cultivated twice), 18% forage crops, and 9% vegetables distributed in the winter and summer seasons. Attainable benefits from intercropping with legumes could be stated. Intercropping faba bean with sugar beet reduces sugar beet infestations with piercing-sucking pest (Hamdany and Aassar 2017). Intercropping onion with sugar beet reduces infestation with tortoise beetle, cotton leafworm, beet moth, and beet fly (Badawy and Shalaby 2015). Intercropping cowpea with sorghum increases its yield by 8% (Abou-Keriasha et al. 2011), and pea intercropped with wheat increased wheat yield by 5% (Zohry et al. 2020b). Maize yield was increased by 10% when intercropped with cowpea (Hamd-Alla et al. 2014). Maize intercropped with watermelon pulp increases pollination in watermelon pulp plants, and consequently increases yield (Sheha et al. 2020). Additionally, intercropping cowpea reduces associated weeds in maize by 7% (Zohry 2005) and in sorghum by 81% (Abou-Keriasha et al. 2011). Furthermore, the quality of summer tomatoes were improved when maize was intercropped with it as maize provides shade on tomato plants (Mohamed et al. 2013).

Table 11 The first suggested crop rotation

	First year	Second year	Third year
One hectare	Faba bean/sugar beet Maize/tomato	Egyptian clover Cotton/wheat	Chickpea/sugar beet Cowpea/sorghum
One hectare	Chickpea/sugar beet Cowpea/sorghum	Faba bean/sugar beet Maize/tomato	Egyptian clover Cotton/wheat
One hectare	Egyptian clover Cotton/wheat	Chickpea/sugar beet Cowpea/sorghum	Faba bean/sugar beet Maize/tomato

– **Winter season crops**

In the first section (Table 11), faba bean (25% of its recommended density) will be intercropped with sugar beet to avoid competition on solar radiation cause by faba bean plants on sugar beet (Zohry and Ouda 2019). In the second section of the rotation, chickpea (25% of its recommended planting density) will be intercropped with sugar beet (Zohry and Ouda 2019). In the third section, Egyptian clover will be cultivated as a temporary mono legume crop (its season length is 75 days). Then, a winter and a summer crops are combined in a relay intercropping system, namely cotton relay intercropped with wheat. In this system, wheat will be cultivation in November in 80% of its planting density in the area assigned to be cultivated with cotton. In March, cotton will be cultivated relay on wheat plants with 100% of its planting density. Cotton seedlings share the period from March to April with wheat plants in their reproductive stage. Wheat harvest occurs in April, and cotton continues to grow until September, where its harvest occurs (Zohry 2005).

– **Summer season crops**

In the first section of the rotation (Table 11), maize (60% of its planting density) will be intercropped with tomato (Mohamed et al. 2013). In the second section, cowpea (50% of its recommended density) could be intercropped with sorghum (Abou-Keriasha et al. 2011).

9.1.2 Second Crop Rotation

The second crop rotation (Table 12) was designed as cereals crop rotation. It included 33% cereal, 25% legumes, 17% sugar beet (cultivated twice), 18% forage crops, and 17% vegetables. Consequently, these rotations could satisfy a part of the demand for food crops.

– **Winter season crops**

In the first section of the rotation (Table 12), onion (33% of its recommended planting density) will be intercropped with sugar beet. In the second section, wheat will be intercropped with sugar beet in 25% of its recommended density (Abou-Elela 2012).

Table 12 The second suggested crop rotation

	First year	Second year	Third year
One hectare	Onion/sugar beet Cowpea/maize	Pea/wheat Maize/watermelon pulp	Wheat/sugar beet Cowpea/sorghum
One hectare	Wheat/sugar beet Cowpea/sorghum	Onion/sugar beet Cowpea/maize	Pea/wheat Maize/watermelon pulp
One hectare	Pea/wheat Maize/watermelon pulp	Wheat/sugar beet Cowpea/sorghum	Onion/sugar beet Cowpea/maize

In the third section, pea will be intercropped with wheat in 50% of its recommended density (Zohry et al. 2020b).

- **Summer season crops**

Cowpea (50% of its recommended density) intercropped with maize will be placed in the first section (Table 12) (Hamd-Alla et al. 2014). Maize (33% of its recommended density) intercropped with watermelon pulp could be implemented in the third section (Sheha et al. 2020) (Table 12).

10 Calculation of Water Requirements of the Suggested Crop Rotations

Drip irrigation with 90% application efficiency is the suggested system to be used in irrigating these crop rotations. Water requirements for the suggested crops in each rotation were calculated using BISm model (Snyder et al. 2004). The weather data for 2019 was obtained from the following website: https://power.larc.nasa.gov/data-access-viewer/. The BISm model calculated evapotranspiration (ETo) using Pennman-Montieth equation (Allen et al. 1998). It uses planting and harvest dates of each crop to determine the values of crop coefficients (Kc) and then determines water consumptive use (WCU) as followed:

$$WCU = ETo * Kc \tag{6}$$

The required water to irrigate each intercropping system were calculated. Then, the total amount of applied water to each rotation was calculated.

The potentially available water from all fish farms in the area in each fish production cycle that could be assigned to crop production was divided by the required irrigation amounts for each crop rotation to determine the potential cultivated area and are presented in Table 13. The table showed that the first suggested crop rotation will require the application of 47,540 m^3/ha. Using the potentially available fish water, this rotation could expand on 19,054 ha. Moreover, the second suggested crop rotation will require 41,380 m^3/ha, and that could allow the expansion on 21,890 ha.

10.1 Interplanting Legume Crops Between Orchard Trees

The third option was to cultivate suitable orchard trees (evergreen or deciduous) and interplant legume crops (winter and summer) between them. The inclusion of legumes in a cropping system improves accessibility of soil nutrients and contributes to the sustainable use of lands. Nitrogen fixation by legumes via a symbiotic relationship with rhizobium bacteria increase mineral soil nitrogen content (Gao et al. 2013).

Table 13 Required irrigation water for the suggested crop rotations and the potential cultivated area

First crop rotation	Required water (m^3/ha)	Second crop rotation	Required water (m^3/ha)
Winter season		*Winter season*	
Faba bean/sugar beet	5,720	Onion/sugar beet	5,720
Chickpea/sugar beet	5,960	Wheat/sugar beet	5,790
Clover	4,460	Pea/wheat	4,260
Wheat	3,680	–	–
Total winter	19,820	Total winter	15,770
Summer season		*Summer season*	
Maize/tomato	10,970	Cowpea/maize	7,380
Cowpea/sorghum	7,780	Cowpea/sorghum	7,780
Cotton	8,970	Maize/watermelon pulp	10,450
Total summer	27,720	Total summer	25,610
Total requirements (m^3/ha)	47,540	Total requirement (m^3/ha)	41,380
Total potential area (ha)	19,054	Total potential area (ha)	21,890

It was documented that the inclusion of legumes in a cropping system could improve soil quality (Espinoza et al. 2015), porosity, and structure (McCallum et al. 2004). Furthermore, more atmospheric nitrogen could be fixed by intercropped legumes than in monoculture, where competition on the fixed nitrogen between the two crops encourages the process (Li et al. 2003). Pea or clover intercropped with grapevine resulted in an increase in soil N, K, and organic matter contents in the soil (Gaser et al. 2017).

Several legume crops are suggested to be interplanted between orchard trees. Faba bean, lupine, and chickpea are the suggested winter legumes, and the suggested summer legumes were cowpea and green bean. If the age of the trees is less than four years, 50% of the recommended planting density of the interplanting crops should be cultivated (Zohry et al. 2020a). For older trees, 30% of the recommended planting density of the interplanting crops should be cultivated to obtain a reasonable yield of both crops (El-Mehy et al. 2017).

Table 14 showed that dividing the total available fish water effluents on the amount of water required for the suggested evergreen cropping system includes legume crops interplanted between orchard trees could result in cultivating 15,676 ha. Alternatively, 10,353 ha of deciduous and legumes cropping systems could be cultivated using the available water amounts.

Table 14 Water requirements for the suggested orchard trees cropping systems and potential cultivated area

Suggested orchard trees	Required irrigation water (m^3/ha)				Cultivated area (ha)
	Orchard trees	Winter legumes[a]	Summer legumes[a]	Total	
Evergreen cropping system					
Citrus	20,839	3,344	7,911	32,095	28,223
Mongo	19,389	3,344	7,911	30,644	29,559
Total evergreen system				57,782	15,676
Deciduous cropping system					
Grape	17,149	3,344	7,911	28,404	31,890
Pomegranate	20,778	3,344	7,911	32,033	28,277
Apricot	21,895	3,344	7,911	33,150	27,324
Total deciduous system				87,492	10,353

[a] Average value of required irrigation water

11 Conclusion

In this chapter, we evaluated the effect of integrating water-smart and soil-smart practices through using fish farm effluent as a source of irrigation water and fertilizer to maximize water and land productivities. The field experiments proved the efficiency of using this appreciated water resource in increasing water and land productivities, which accomplish the first pillar of climate-smart agriculture.

Furthermore, based on the field experiments, the modeling analysis used in this study gave a general idea about the potential use of fish farms effluents. Using fish farm effluent could attain improvements in the management of such valuable water resources. We demonstrated that these large amounts of fish farm effluents could be invested in irrigating a large area with field crops or orchards to increase food production. As a sequence, fertilizer use in crop production will be minimized and deterioration of freshwater resources will be prevented.

References

AbdEl Magid HAA, Hala HAA, Mohamed AMM (2018) Economic study the efficiency of water resource usage (Case Study). Alex J Agric Sci 63(3):149–155

Abdel-Aziz RA, Salem SE, Al-Bitar L (2008) Effect of intercropping cover crops on citrus orchards growth and fruiting under Toshka conditions. J Agric Veterinary Sci Qassim Univ 1(2):101–110

Abdelraouf RE, Hoballah EMA (2014) Impact of irrigation systems, fertigation rates and using drainage water of fish farms in irrigation of potato under arid regions conditions. Int J Sci Res Agric Sci 1(5):67–79

Abdelraouf RE, Ragab R (2017) The benefit of using drainage water of fish farms for irrigation: field and modelling study using the SaltMed model. Irrig Drain 66:758–772

Abou-Elela AM (2012) Effect of intercropping system and sowing dates of wheat intercropping with sugar beet. J Plant Prod 3(12):3101–3116

Abou-Keriasha MA, Ibrahim ST, Mohamadain EEA (2011) Effect of cowpea intercropping date in maize and sorghum fields on productivity and infestation of weed. Egypt J Agron 33(1):35–49

Allen RG, Pereira LS, Raes D, Smith M (1998) Crop evapotranspiration: guideline for computing crop water requirements. FAO No56

Badawy ShAE, Shalaby GA (2015) Effect of intercropping of sugar beet with onion and garlic on insect infestation, sugar beet yield and economics. J Plant Prod Mansoura Univ 6:903–914

Bakshi P, Bhushan A, Bali K, Kour K (2019) Intercropping in fruit orchards: a way forward for doubling the farmer's income. 11(23):9274–9276

Castro RS, Azevedo CMB, Bezerra-Net F (2006) Increasing yield of cherry tomato using fish effluent as irrigation water in Northeast Brazil. Sci Hortic 110(1):44–506

Dakkak A (2016) Egypt's water crisis—recipe for disaster. https://www.ecomena.org/egypt-water/

Direcenmattos Jr, Yamane DR, Boaretto RRM, Zambrosi FCB, Quaggio JA (2010) Root development of young citrus trees in soils fertilized with phosphorus. In: 19th world congress of soil science. Soil solutions for changing world 1–6 August 2010, Brisbane, Australia, pp 197–203

El Bedawy R (2014) Water resources management: alarming crisis for Egypt. J Manag Sustain 4(3):108–124

El-Mehy A, El-Badawy HEM (2017) Evaluation of intercropping corn, soybean and cowpea with Washington navel orange orchard between different N fertilizer levels. Middle East J Agric 06(02):513–533

El-Mehy AA, Taha A, M, Abd-Allah AAMM (2018) Maximizing land and water productivity by intercropping sunflower with peanut under sprinkler irrigation. Alex Sci Exch J 39(1):144–159

Elnwishy N, Salh M, Zalat S (2006) Combating desertification through fish farming: the future of drylands. In: Proceedings of the international scientific conference on desertification and drylands research, Tunisia 19–21, June UNESCO

Espinoza S, Ovalle C, Zagal E, Matus I, del Pozo A (2015) Contribution of legumes to the availability of soil nitrogen and its uptake by wheat in Mediterranean environments of central Chile. Chilean J Agric Res 75:111–121

Fedoroff NDS, Battisti RN, Beachy PJM, Cooper DA, Fischhoff CN, Hodges VC, Knauf D, Lobell BJ, Mazur D, Molden MP, Reynolds PC, Ronald MW, Rosegrant PA, Sanchez AV, Zhu JK (2010) Radically rethinking agriculture for the 21st century. Sci 327:833–834

Gad A, El-Zeiny A (2016) Spatial analysis for sustainable development of El-Fayoum and Wadi El-Natrun desert depressions, Egypt with the aid of remote sensing and GIS. J Geog Environ Earth Sci Int 8(3):1–18

Gao L, Xu H, Bi H, Xi W, Bao B, Wang X, Bi C, Chang Y (2013) Intercropping competition between apple trees and crops in agroforestry systems on the Loess Plateau of China. PLoS ONE 8(7):e70739. https://doi.org/10.1371/journal.pone.0070739

Gaser ASA, Thoraua SA, Abo El-Wafa MA, El-Kenawy SA (2017) Effect of intercropping of peas and clover corps on growth, productivity and soil characteristics of flame seedless and Thompson seedless grapevine cultivars. Egypt J Hort 44(2):183–197

Gilbert GA, Knight JD, Vance CP, Allan DL (1999) Acid phosphatase activity in phosphorus-deficient white lupine roots. Plant Cell Environ 22:801–810

Hamd-Alla WA, Shalaby EM, Dawood RA, Zohry AA (2014) Effect of cowpea (Vigan sinensis L.) with maize (Zea mays L.) intercropping on yield and its components. Int Scho Sci Res Innov 8(11):1170–1176

Hamdany MKH, Aassar MR (2017) Effect of intercropping three faba bean varieties with sugar beet plants on piercing sucking insect pests and associated natural enemies under ridge space and seedling rates in relation crop yield. Egypt Acad J Bio Sci 10(6):57–77

Haquea MM, Belton B, Mehedi AMD, Golzar AA, Rushna AMd (2016) Reuse of fish farm sediment as fertilizer for fodder grass production in Bangladesh: potential for sustainable intensification and improved nutrition. Agric Ecosyst Environ 216:226–236

Hauggaard-Nielsen H, Mundus S, Jensen ES (2009) Nitrogen dynamics following grain legumes and subsequent catch crops and the effects on succeeding cereal crops. Nutr Cycl Agroecosyst 84(3):281–291. https://doi.org/10.1007/s10705-008-9242-7

Isitekhale HHE, Adamu B (2016) Effects of effluents on soil chemical properties in forest derived savanna transition. IOSR J Environ Sci Toxicol Food Technol 10(6)11:30–34

Kestha MM, El-Baz MG (2004) Studies on sunflower-tomato intercropping. In: Proceedings of the 16th international sunflower conference, Fargo, ND USA, pp 514–518

Li L, Zhang F, Li X, Christie P, Sun J, Yang S, Tang C (2003) Interspecific facilitation of nutrient uptake by interplanted maize and faba bean. Nutr Cycl Agroecosyst 65:61–71

Maillard P, Santos NAP (2008) A spatialstatistical approach for modeling the effect of non-point source pollution on different water quality parameters in the Velhas river watershed-Brazil. J Environ Manag 86(1):158–170

Mao LL, Zhang LZ, Li WW, Werf WVD, Sun JH, Spiertz H, Li L (2012) Yield advantage and water saving in maize/pea intercrop. Field Crop Res 138:11–20

McCallum MH, Kirkegaard JA, Green T, Cresswell HP, Davies SL, Angus JF (2004) Improved subsoil macro-porosity following perennial pastures. Aust J Exper Agric 44:299–307

Mead R, Willey RW (1980) The concept of a "land equivalent ratio" and advantages in yields from intercropping. Exper Agric 16:217–228

Mohamed W, Ahmed NR, Abd El-Hakim WM (2013) Effect of intercropping dates of sowing and N fertilizers on growth and yield of maize and tomato. Egypt J Appl Sci 28(12B):625–644

Morgado LN, Carvalho CF, Souza B, Santana MP (2002) Fauna of bees (Hymenoptera: Apoidea) on sunflower flowers, Helianthus annuus L., in Lavras - MG. Ciência Agrotecnologia 26:1167–1177

Ojobor SA, Tobih FO (2015) Effects of fish farm effluent and inorganic fertilizer on Amaranthus yield and soil chemical properties in Asaba, Delts State Nigeria. J Agric Environ Sci 4(1):237–244

Ouda S, Hefny Y, Abdel-Wahab S, Abdel-Wahab T (2018) Intercropping patterns of sunflower with peanut under different irrigation regimes and potassium fertilizer levels. Egypt J Agron. In: The 5th international conference for crops science. SI:85–104

Said AS, Zohry AA, Ouda S (2016) Unconventional solution to increase crop production under water scarcity. In: Major crops and water scarcity in Egypt. Springer Publishing House, pp 99–114. ISBN 978-3-319-21770-3

Selim MAF, Shams AS (2019) Maximizing efficiency of land and water utilization and profitability of interplanting maize with mandarin trees using irrigation with fish waste water between sandy soil and drip irrigation conditions. Middle East J Agric Res 8(04):1240–1252

Selim M, Hefny Y, Abdel-Wahab E, Mohamed M (2020a) Interplanting some soybean cultivars with mandarin trees in sandy soil. Agric Sci 11:88–110. https://doi.org/10.4236/as.2020.111006

Selim M, Hefny Y, Mostafa MM (2020b) Interplanting soybean with three species of orchard trees under two soybean plant distributions in sandy soils. Acta Sci Agric 4(8):21–36

Sheha AM, El-Shahed HM, Diab AHM (2020) Response of watermelon pulp (*Citrullus colocynthoides)* to intercropped maize (*Zea mays L.*) with plants distributions under different nitrogen fertilization levels. J Agric Sci12(8):212–225

Snyder RL, Orang M, Bali K, Eching S (2004) Basic irrigation scheduling (BISm). http://www.waterplan.water.ca.gov/landwateruse/wateruse/Ag/CUP/Californi/Climate_Data_010804.xls

Steingrobe B (2005) Root turnover of faba beans (Vicia faba L.) and its interaction with P and K supply. J Plant Nutr Soil Sci 168:364–371

van der Heijden PGM (2012) Water use at integrated aquaculture agriculture farms. Experiences with limited water resources in Egypt. Wageningen University and Research Centre for Development Innovation

Watt M, Evans JR (2003) Phosphorus acquisition from soil by white lupin (Lupinus albusL.) and soybean (Glycine max L.), species with contrasting root development. Plant Soil 248:271–283

Weisskopf L, Heller S, Eberl L (2011) Burkholderia species are major inhabitants of white lupine cluster roots. Appl Environ Microbiol 77(21):7715–7720. https://doi.org/10.1128/AEM.05845-11

Zieman DA, Walsh WA, Saphore, Berneth KF (2007) A survey of water qualities characteristics of effluent from Hawaian Facilities. J World Aqua Soc 23(3):180–191

Zhang Q, Li ZW, Zeng GM, Li JB, Fang Y, Yuan QS (2009) Assessment of surface water quality using multivariate statistical techniques in red soil hilly region: a case study of Xiangjiang watershed, China. Environ Monitor Assess 152:123–131

Zohry AA (2005) Effect of relaying cotton on some crops under bio-mineral N fertilization rates on yield and yield components. Annals of Agric Sci 431:89–103

Zohry A, Ouda S (2018) Suggested crop rotations to increase food security in Egypt. In: Crop rotation: an approach to secure future food. Springer Publishing House. ISBN 978-3-030-05350-5

Zohry A, Ouda S (2019) Intercropping systems for sugar beet to improve its land and water productivity. J. Soil Crop 29(2):218–226

Zohry AA, Ouda S, Hamd-Alla W, Shalaby E (2017) Evaluation of different crop sequences for wheat and maize in sandy soil. Acta Agric Slov 109(2):383–392

Zohry A, Hefny Y, Ouda S (2020a) Interplanting four legume crops under orange trees using different irrigation water and fertilizer sources in sandy soil. In: Proceedings of the 16th international conference on crop science, 16th of October, Cairo, Egypt, pp 341–346

Zohry A, Ouda S, Abdel-Wahab T (2020b) Sustainable intensive cropping to reduce irrigation-induced erosion: Intercropping systems under surface irrigation practice. Morocco J Agric Sci 1(2):63–71

Chapter 4
Integration Between Crop-Smart, Water-Smart and Soil-Smart Practices

Abd El-Hafeez Zohry and Samiha Ouda

Abstract To sustainably increase food production, crop-smart, water-smart and soil-smart practices should be integrated. Implementing intercropping systems can attain the integration of the three practices because it have many advantages. It attains higher crops productivity and increase water productivity. It can attain soil conservation, as well as control the insects, diseases and weeds. It can also modify the microclimate of the intercrops to increase its yield. Other crop-smart practices are the cultivating three crops in one year and implementing crop rotations. Crop rotation also integrates the crop-smart, water-smart and soil-smart practices. Examples of crop rotations implemented in Egypt under rain-fed conditions and irrigated agriculture were reviewed, namely cotton rotation, cereal crops rotations, intensive-cropping rotations, crop rotations to defeat diseases and weeds and water saving crop rotations.

Keywords Intercropping systems · Soil conservation · Water conservation · Sequence of cultivated crops · Crop rotations

1 Introduction

Weather, soil, and water are the natural resources that control food production in a certain area. Crops growing season, and the required irrigation water for crops are dependent on weather conditions. Soils are the medium, in which nearly all food-producing plants grow. Furthermore, water is considered as the most critical natural resource for sustainable development in most areas of the world. Sustainable use of soil and water can guarantee the continuity of food production. Production of food requires the integration between crops species, soils, and water in a sustainable manner to produce high amounts to feed the growing population. This integration

A. E.-H. Zohry (✉)
Crops Intesefication Research Department, Field Crops Research Institute; Agricultural Research Center, Giza, Egypt

S. Ouda
Water Requirements and Field Irrigation Research Department, Soils, Water and Environment Research Institute; Agricultural Research Center, Giza, Egypt

S. Ouda and A. E.-H. Zohry, *Climate-Smart Agriculture*,
https://doi.org/10.1007/978-3-030-93111-7_4

realized the first pillar of climate-smart practices that addresses food security and attained one of UN SDGs goals.

Several practices can be conducted to accomplish sustainable food production, namely implementing intercropping systems, suitable sequence of the cultivated crops, and crop rotations. Many benefits, other than increasing food availability could be realized from these practices. Improve in soil fertility can be attained from including legumes in the intercropping system (Zohry et al. 2017a; Ouda et al. 2020), or from its inclusion in a crop sequence (Hamd-Alla et al. 2014). Some intercropping systems can mitigate biotic (Hamd-Alla et al. 2014; Hamdany and Aassar 2017) and abiotic stresses (Abd El-Zaher et al. 2013). Furthermore, Crop rotations have an important role in the sustainable use of natural resources. In addition to increase food availability and improve soil fertility, it can control weeds, insects and diseases. Crop rotations using cover crops could significantly improve soil organic carbon, N, and water cycles (Camarotto et al. 2018). To improve efficiency of a crop rotation, it should include both intercropping systems and appropriately sequenced crops. These practices can be implemented in both irrigated and rain fed area to attain the benefits of climate-smart agriculture. Under rain-fed conditions, soil moisture shortage is the factor controlling crop production, therefore soil and water conservation methods should be implemented (Kiboi et al. 2017). Wolka et al. (2018) indicated that soil and water conservation techniques improve soil fertility by reducing soil and nutrient loss. In this chapter, crop-smart practices and its integration with water-smart and soil-smart practices and its role in attaining the sustainable use of natural resources for food production were discussed.

2 Intercropping Systems

Intercropping offers a potential benefit over monoculture cultivation for yield increase through more effective use of resources, including water, nutrients, and solar energy (Nasri et al. 2014). It is considered a successful avenue to increase the cropped area without altering the area cultivated by the main crops in winter or summer seasons (Kamel et al. 2010). Katyayan (2005) defined intercropping as a system of management of crops which involves growing two or more dissimilar crop species simultaneously in distinct row combination on the same piece of land. It is also a planting system involve cultivation of two crops (main and secondary crops) resulted in increasing light interception, root contact with more soil, microbial activity and it deterrent to pests and weeds (Altieri 1999). This practice could be used as a way to improve soil fertility, increase land productivity, as well as increase water productivity as a result of using less water to irrigate two crops (Andersen 2005).

The conventional ways of intensifying crop production are vertical and horizontal expansions, however intercropping offers two additional dimensions, time and space (Zohry and Ouda 2019). Francis (1986) stated that the intensification of land and resources use in space dimension is an important aspect of intercropping, where it enhance the efficient use of light as two or more species occupying the same land during a significant part of their growing seasons and have different pattern of foliage display. Furthermore, different rooting patterns can explore greater and different soil volume and depth (Francis 1986). On the other hand, intercropping systems need comparatively fewer energy inputs like fertilizer, plant protection chemical requirements. The diversification of crops is greater creating functional diversity resulting in less pest-disease incidence (Gitari et al. 2018). Several advantages can be attained from implementing intercropping systems as followed.

2.1 Intercropping Attains Yield Advantage

Implementing intercropping systems can increase food availability and security (Ouma and Jeruto 2010). The most common reason for the adoption of intercropping systems technique is yield advantage, which is explained by the greater resource depletion by intercrops than monocultures (Hauggaard-Nielsen et al. 2009). The efficiency of the intercropping is directly depends on proper management of the factors of production, such as the spatial arrangement of crops and planting density, where it can reduce the competition for resources and increase the efficiency of the system (Porto et al. 2011). These factors, when properly managed, can bring ecological and economic benefits, as a result of increasing production when compared to monoculture (Batista et al. 2016). Thus, intercropping increases unit land productivity (harvest two types of crops from the same area), and increase farmers income (reduce risks from crop failure) (Anderson 2007).

3 Land Equivalent Ratio

Evaluation of the efficiency of an intercropping system to attain yield advantage over the sole planting of each crop can be done by several indices. The most important one is land equivalent ratio (LER). LER is the ratio of the area under sole cropping to the area under interplanting needed to give an equal amounts of yield at the same management level. It is the sum of the fractions of the interplanted yields divided by the sole crop yields (Mead and Willey 1980). LER is calculated as follows:

$$\begin{aligned} L_A &= Y_{IA}/Y_{AA} \\ L_B &= Y_{IB}/Y_{BB} \\ LER_{total} &= L_A + L_B \end{aligned}$$

where L_A and L_B = Relative yield of crop A and B, Y_{IA} and Y_{IB} = intercropped yield of crop A and crop B, Y_{AA} and Y_{BB} = Pure stand yield of crop A and crop B, respectively.

Table 1 Values of land equivalent ratio (LER) presented in the literature for several intercropping systems in Egypt

Intercropping system	Planting density (100%)		LER	References
	Secondary	Main		
Peanut/maize	50	100	1.4	Shams et al. (2012)
Faba bean/onion	50	75	1.6	Abou-Keriasha et al. (2013)
Soybean/maize	50	50	1.4	Lamlom et al. (2015)
Soybean/sunflower	50	100	1.3	Abdel-Wahab and El Manzlawy (2016a)
Cowpea/maize	50	100	1.5	Abdel-Wahab et al. (2016)
Faba bean/wheat	50	67	1.3	Abdel-Wahab and El Manzlawy (2016b)
Faba bean/sugar beat	50	100	1.5	El-Mehy et al. (2020)
Maize/tomato	70	100	1.8	El-Mehy and Mohamed (2018)
Sunflower/peanut	50	100	1.4	Ouda et al. (2018)
Faba bean/sugar beet	33	100	1.3	Zohry and Ouda (2019)
Onion/sugar beet	33	100	1.6	Zohry and Ouda (2019)
Cowpea/sunflower	50	100	1.5	Zohry et al. (2020a)

If the LER > 1, it suggests that the land utilization of intercropping is higher than that of monoculture. If LER < 1, it shows that land utilization of intercropping is lower than that of monoculture.

It was reported by several authors in Egypt that LER values for different intercropping systems were higher than under monoculture cultivation (Table 1). The table showed that the planting density of the main crop in most of the intercropping systems was 100% and the secondary crop was less than 100%. It also showed that LER value could reach as high as 1.8 in the case of intercropping maize on summer tomato (El-Mehy and Mohamed 2018). Consequently, the income of the farmers could be increase by implementing intercropping systems, compare to sole cultivation of each crop included in the intercropping system.

3.1 *Higher Water Productivity Under Intercropping Systems*

Intercropping system makes full use applied irrigation water (Qin et al. 2013). It increases water productivity as a result of using amount of irrigation water applied to one crop to irrigate two crops under intercropping system (Anderson 2007). Yang et al. (2011) reported that intercropping has been observed to improve productivity per unit area through efficient and complementary use of water. Coll et al. (2012) indicated that the great yields attained by the intercrops are only as a consequence of low water losses. Furthermore, Miao et al. (2016) found that actual evapotranspiration, irrigation water use, crop transpiration, and groundwater contribution of intercropping systems were larger than those of the sole crops, which led to significantly higher land and water equivalent ratios of intercropping than those of single crops. In intercropping, the maximum ground area is covered; hence there will be a minimum chance of runoff (Maitra 2020). For example, sorghum intercropped with cowpea system reduced runoff by 20–30% compared to sorghum sole planting and by 45–55% compared to cowpea monoculture (Lithourgidis 2011).

4 Water Equivalent Ratio (WER)

To judge the efficiency of water utilization by intercropping system, water equivalent ratio (WER) is used. WER reflects water use advantage of intercropping systems (Chimonyo et al. 2016). A higher than 1 value of WER of intercropping systems indicates higher water utilization efficiency of the intercropped crops than that of monoculture (Feng et al. 2016). It is defined as the total water needed in sole crops to produce the equivalent amount of crop yield on a unit area of intercrop (Mao et al. 2012). WER was calculated for each crop sequence interplanted under orange trees as follows:

$$\mathrm{WER} = \frac{\left(\frac{Y_{\mathrm{int,A}}}{WU_{\mathrm{int,A}}}\right)}{\left(\frac{Y_{\mathrm{mono,A}}}{WU_{\mathrm{mono,A}}}\right)} + \frac{\left(\frac{Y_{\mathrm{int,B}}}{WU_{\mathrm{int,B}}}\right)}{\left(\frac{Y_{\mathrm{mono,B}}}{WU_{\mathrm{mono,B}}}\right)}$$

where $Y_{\mathrm{int,A}}$, and $Y_{\mathrm{int,B}}$ are the yield of intercropped crops A and B, respectively. $WU_{\mathrm{int,A}}$, and $WU_{\mathrm{int,B}}$ are water consumptive use by the intercropped crops A and B, respectively. $Y_{\mathrm{mono,A}}$, and $Y_{\mathrm{mono,B}}$ are the yield of mono crops A and B, respectively. $WU_{\mathrm{mono,A}}$, and $WU_{\mathrm{mono,B}}$ are water consumptive use by mono crops A and B, respectively.

If the WER > 1, it suggests that the water utilization of intercropping is higher than that of monoculture and that imply advantage in implemented intercropping system. If WER < 1, it shows that water utilization of intercropping is lower than that of monoculture and that imply disadvantage. Table 2 presented published values of WER of selected intercropping systems implemented in Egypt.

4.1 *Intercropping Realizes Soil Conservation*

Intercropping is an excellent practice for controlling soil erosion and sustaining crop production (Dwivedi et al. 2015). Deep roots of legume crops penetrate far into the soil and use moisture and nutrients from deeper soil layers, whereas shallow roots of cereal crops fix the soil at the surface and thereby help to reduce erosion

Table 2 Values of water equivalent ratio (WER) presented in the literature of several intercropping systems in Egypt

Intercropping system	Planting density (100%)		WER	References
	Secondary	Main		
Sunflower/peanut	50	100	1.3	Ouda et al. (2018)
Maize/mandarin trees	50	100	1.6	Selim and Shams (2019)
Faba bean/orange trees	50	100	1.7	Zohry et al. (2020b)
Lupine/orange trees	50	100	1.8	Zohry et al. (2020b)
Faba bean/sugar beet	25	100	1.3	El-Mehy et al. (2020)
Chickpea/sugar beet	33	100	1.3	Zohry and Ouda (2019)
Faba bean/sugar beet	33	100	1.3	Zohry and Ouda (2019)
Onion/sugar beet	33	100	1.5	Zohry and Ouda (2019)

(Machado 2009). In intercropping, the maximum ground area is covered; hence there will be a minimum chance of soil erosion and nutrients loss (Nyawade et al. 2019). In low fertile soils, legume and cereal intercropping systems can increase soil fertility via raising its organic content and available nitrogen fixed by the legume (Megawer et al. 2010). Some legume crops are better in maintaining crop cover than the others. Kariaga (2004) compared between cowpea intercropped with maize system and bean intercropped with maize system regarding soil erosion and found that cowpea preformed, as the best cover crop, compared to bean in reducing soil erosion. Additionally, soil loss was reduced with intercropping cowpea with sorghum by more than 50%, compared to the monoculture of each crop (Lithourgidis 2011).

The cereal-legume combination of intercropping is known to enable long term immobilization of N (Iqbal et al. 2018). Nutrient balance studies indicated that cereal-legume intercropping enhanced N fertility of the soil. Among legumes, groundnut in combination with maize added N to the soil because of above- and below-ground architecture of groundnut and it ability to provide more soil coverage (Choudhary and Choudhury 2016). In Egypt, Ouda et al. (2020) indicated that soil nitrogen percentage was increased by 75% when cowpea was intercropped with maize, compared to the percentage existed after sole maize cultivation. Zohry et al. (2020b) stated that an increase in soil organic matter was observed when cowpea was intercropped with sunflower by 8% and by 9% when pea intercropped with wheat.

The interactions among microbes, nutrients and enzymes in intercropping systems leads to an increase or decrease in microbe quantity and enzyme activity, contributing to the improvement of the soil micro-ecological environment (Zhou et al. 2019). Intercropping peanut with sugarcane changes the bacterial diversity of soils and decrease disease rates (Tang et al. 2021).

4.2 Intercropping Control Crop Predators' Infestation

Intercropping systems can control the insects, diseases and weed population dynamics. Above ground, diversity is caused by the inclusion of two or more crops in intercropping. These crop mixtures attract pollinating bees and other predators which has a significant impact on production enhancement and insect population dynamics (Nicholls and Altieri 2013). When cowpea was grown in an intercropping system with cotton, the cotton sucking pest population was reduced (Chikte et al. 2008). Under intercropping faba bean with sugar beet, the rate of infestation with sucking pests was lower than its value in the sole sugar beet planting (El-Fakharany et al. 2012). Growing of sorghum as an intercrop significantly reduced the incidence of sucking pests of groundnut (Parthiban et al. 2016).

Weeds population was also reduced under intercropping systems. Abou-Keriasha et al. (2011) indicated that cowpea intercropped with sorghum system caused a reduction in the associated weeds with sorghum by 81%, compared to sole planting. Similar trend was found when cowpea was interplanted with maize (Silva et al. 2009).

4.3 Intercropping Modifies the Microclimate of the Associated Crops

Intercropping practice could modify the microclimate by reducing light intensity, air temperature, desiccating wind and other climatic components. It can increase the amount of light intercepted to crops in a unit planting area, which improve radiation use efficiency, crop dry matter production and yield (Tsubo and Walker 2002). Plant density and row orientation in an intercropping system plays an important role in the received irradiance and accordingly affects canopy maximum temperatures (Jaya et al. 2001).

Grain sorghum intercropped with tomato as the shade crop yielded more than pure stand tomato with little loss of sorghum yield where sorghum reduces air temperatures surrounding intercropped tomato canopy to 5–7 °C, as compared with the pure stand tomato (Kamel et al. 2004). Shading provided from sunflower plants intercropped with summer tomato plants protect their fruits from sun damage in July and August (Kestha and El-Baz 2004). Intercropping faba bean (Ibrahim et al. 2010) or wheat (Abd El-Zaher et al. 2013) with tomato in the winter reduces the damage of low temperature and provide warmth to tomato plants, and that resulted in extension in its growing season for two months. Maize intercropped with peanut system enhances strong light utilization ability and leads to higher efficiency of soil nutrients than monoculture planting patterns (Zhang et al. 2019).

In addition, interplanting sunflower between orange trees with low distance between sunflower plants protected sunflower plants from winds allowing more visitations of bees, which are the most important insects in sunflower pollination process (Morgado et al. 2002).

5 The Sequence of the Cultivated Crops

Implementation of intensive cropping to increase food production demand the sustainable use of water and land resources. Accordingly, the appropriate crop sequence that maximize the use of water and land resources should be studied intensively to attain the objectives of the first pillar of climate-smart agriculture. The sequence of crops planted in a piece of lands play an important role in increasing the productivity of the unit lands and improve soil fertility which consequently increase water use efficiency (FAO 2003). Well sequenced crops improve soil quality and crops production (Hamd-Alla et al. 2014).

Legume crops are good candidates to be included in a crop sequence because of its role in increasing soil fertility (Sheha et al. 2014), raising the organic content and increasing available nitrogen contents in the soil (Singh et al 2003), which positively affects the following crops (Kumpawat and Rathore 2003). Stone and Silveira (2001) indicated that the sequence of soybean then wheat resulted in the highest soil microporosity values and the lowest soil macro-porosity values, whereas the sequence of upland rice associated and common bean showed the highest soil macro-porosity value and the lowest micro porosity value.

Several studied were done in Egypt to determine the appropriate crop sequence of field crops to sustainably increase its productivity. Lamlom et al. (2015) indicated that when clover in the winter preceded soybean intercropped with maize in the summer, soybean and maize yield were increased by 9 and 12%, respectively, compared to wheat proceeding it. Furthermore, Zohry et al. (2017a) indicated that inclusion of short-season clover (70 days) between maize and wheat cultivation resulted in yield increase in maize and wheat in the following year by 16 and 47%, respectively. Abou-Khadra et al. (2018) concluded that inclusion short-season clover between rice and wheat or maize and application of ascobien acid could be an alternative way to decrease mineral N inputs to wheat by about 25% of the recommended dose. Ouda et al. (2020) stated that wheat yield was increased by 30%, when short-season clover was cultivated between maize and wheat, and by 55% when wheat was preceded by cowpea then short-season clover, compared to cultivation of maize followed by wheat as a crop sequence.

Other studies have been done to determine the effect of crop sequence on soil fertility status. Abbady et al. (2016) reported that intercropping pea with wheat followed by intercropped cowpea with sunflower increased soil organic matter by 31%. Whereas, Zohry et al. (2017b) indicated that soil organic matter was increased by 11% after the above sequence. Ouda et al. (2020) reported that the value of soil nitrogen percentage was increased by 113 and 150%, when short-season clover was cultivated between maize and wheat, and when wheat was preceded by cowpea then short-season clover was cultivated, respectively.

6 Crop Rotation

Arable crops are grown by farmers in patterns according to time and space. A crop rotation defined by Thenail et al. (2009) as "growing crops in a recurring sequence on the same field". Crop rotation is the production of different economically important plant species in recurrent succession on a particular field or group of fields (Bruns 2012). The benefit of crop rotations is well accepted in traditional farming approaches, as well as supported by scientific knowledge (Bennett et al. 2012). Crop rotation means changing the type of crop grown on a particular piece of land from year to year. It is one of most effective agricultural control strategies. It involves arrangement of crops planted on same field, where the succeeding crops should belong to different families (Huang et al. 2001).

In a crop rotation, alternate deep and shallow rooting crops improve soil structure and aeration. In addition, alternate crops with large and small root biomass results in increase the organic matter remaining in the soil for soil microbial and macro faunal populations (Bennett et al. 2012). Furthermore, rotate N2-fixing and N-demanding crops attempt to meet farm's N demands from within the system (Malik 2010).

One of the most important benefits of applying crop rotation was to prevent the spread of pests, weeds and diseases through rotating the crops, where inclusion of breaking phases decreases its population dynamics (Kirkegaard et al. 2008). An effective crop rotation deployment might limit, and in some cropping years even eliminate, the need for herbicide applications to control the dominant weed species (Gonzalez-Diaz et al. 2012). Changes in crop rotation and herbicide use could change the weed seed banks in arable soils (Squire et al. 2000).

Crop rotation reduces risk of weather damage, thus reduces yield losses and that will increase net profit of farmers (Kaye et al. 2007). It could also improve grain quality, and reduced grain yield variability (Varvel 2000). Crop rotation increases water use efficiency and improve crop nutrients use efficiency (Tanaka et al. 2005). Furthermore, several practices can be use within the rotation to reduce the applied irrigation water to the cultivated crops, namely land leveling, and cultivation on raised beds (Ouda and Zohry 2018). Implementing intercropping systems within the crop rotation is known to increase water productivity, as a result in saving a large amount of irrigation water.

Crop rotation provides integration between crop-smart, water-smart and soil-smart practices. It plays an important role in increasing number of cultivated crops per year, and, consequently increase the total production, which can reduce food gaps. Diverse crop rotations provide a yield-enhancing, compared to less diverse rotations (Berzsenyi et al. 2000). Furthermore, diversifying cropping systems with annual pulse crops, such as pea, lentil, and chickpea could increase the systems' productivity, while decreasing the environmental impact (Gan et al. 2015).

7 Crop Rotations in Egypt

In Egypt, several studies were done on implementing crop rotations to attain the objective of the integration of crop-smart, soil-smart and water-smart practices in both rain fed areas and under irrigated agriculture.

– **Crop rotations under rain fed**

In Egypt, under rain fed agriculture in North Egypt, Khalifa et al. (2004) reported an increase in barley yield under two-year crop rotation, when it was followed by pea or lentil, namely by 113 and 104%, respectively compared to barley monoculture. These results implied that continuation of barley cultivation exhausted the soil and depleted nutrients from it and including legumes in the crop rotation is a recommended procedure for sustaining soil productivity and improving barley productivity under rain fed conditions.

Furthermore, Shams and Kamel (2014) conducted monoculture rotation contained either barley or wheat and compare it with two-year rotation, where barley or wheat rotated with lentil or pea. Their results revealed that the yield of either barley or wheat was drastically decreased in monoculture by decreased by 34 and 28%, respectively. Whereas, barley and wheat yield reductions in the two-year rotation were lower, namely 13 and 9%, respectively. Although the two cereal crops were rotated with the two legume crops, the reduction in the cereal crops was due to lower incidence of rain in the second year of the rotation.

Five two-year crop rotations were implemented in North Egypt under rain fed condition to assess soil fertility build up, as result of implementing these rotations. These rotations were barley-fallow, wheat-wheat, barley-barley, barley-pea, and barley-lentil. The results indicated that the highest soil organic matter was obtained when barley-pea rotation was implemented, where organic matter was increased by 47% after the harvest. The values of N, P, and K were increased by 73, 300, and 19%, respectively under barley-pea rotation (Abd El-Hadi et al. 2002).

– **Crop rotations under irrigated agriculture**

7.1 Cotton Rotations

In Egypt, several studies were done for to assess cotton yield under different rotations. Abou-Kersha (1998) indicated that the yield of cotton rely intercropped on faba bean was higher by 5 under two-year crop rotation and by 6% under three-year rotation, compared to the yield resulted from cotton monoculture.

Similar results were found by Toaima (2007) for cotton, where its yield was increased by 11 and 24% under the two-year and three-year rotations, respectively compared to cotton cultivated every year. Furthermore, in the two-year and three-year crop rotation, the associated weeds with cotton were decreased by 31 and 48%, compared to the one-year crop rotation.

Similarly, Farghly and Zohry (2002) evaluated the effect of the preceding crop sequences on cotton yield in a three-year crop rotation. Four crop rotations were implemented, where in each rotation different crop sequences preceded cotton as follows: wheat/maize, wheat/soybean, faba bean/maize and faba bean/soybean. In the end of the first rotation, where wheat/maize preceded cotton, soil available N, P, and K was reduced by 31, 5 and 8%, respectively and cotton yield was the lowest.

In the end of the second rotation, where wheat/soybean preceded cotton, the soil available N, P, and K in the soil was reduced by 13, 5, and 5%, respectively. In the end of the third rotation, where faba bean/maize preceded cotton, the soil available N, P, and K were increased by 38, 34, and 2%, respectively. Whereas, in the end of the fourth rotation, where faba bean/soybean preceded cotton, the soil available N, P, and K were increased by 50, 64 and 5%, respectively and cotton yield was 14% higher, than its yield attained in the first rotation.

7.2 *Intensive-Cropping Rotations*

Additionally, Abou-Keriasha et al. (2012) implemented three cotton rotations, three years each, namely mono crops, the second crop rotation and the third crop rotation. They evaluated the success of these crop rotations by two measurements, namely intensification index and cereal units as followed.

7.2.1 Intensification Index

The intensification index of crop rotation, which represent the sum of the cropped area divided by the land area multiplied by 100.

$$\textit{Intensification index} = \sum(\textit{Cropped area/Land area}) * 100$$

7.2.2 Cereal Units (CU)

To calculate cereal units, the yield of each crop in the rotation was changed to units of cereal. It has been used as a common denominator in German agricultural statistics for decades and is mainly based on the nutritional value for livestock. It is also an appropriate unit for the description of agricultural products (Brankatschk and Finkbeiner 2014). Furthermore, Macák et al. (2015) used cereal units to evaluate productivity of different crop rotations. According to Brockhaus (1962), it allows judicious comparison between 100 kg of each crop. Accordingly, 100 kg of either wheat or maize = 1.00 CU, 100 kg of clover and cowpea = 1.14 and 1.12 CU,

Year 1	Year 2	Year 3
Clover (one cut) Cotton	Faba bean Maize	Wheat Maize
Wheat Maize	Clover (one cut) Cotton	Faba bean Maize
Faba bean Maize	Wheat Maize	Clover (one cut) Cotton

Fig. 1 The prevailing cotton crop rotation (first rotation)

respectively. 100 kg of faba bean = 1.20 CU, 100 kg of sesame = 2.00 CU, 100 kg of onion = 0.30 CU, and 100 kg of cotton = 5.55 CU.

The first crop rotation is the prevailing rotation in the surrounding areas, where crops are cultivated on rows under surface irrigation (Fig. 1).

The second crop rotation was a low-intensive crop rotation (contained mono crops, intercropping systems and three-crop sequence, mainly legumes) and crops are grown on raised beds (Fig. 2). In the first section, cotton was relayed intercropped with onion (Zohry 2005). After the harvest of onion, sesame intercropped with cotton and replaced onion. In this system, sesame was cultivated on one row above the raised beds and continue to grow until September, where cotton harvest occur. Thus, the above intensive system started in November by the cultivation of onion and ended in September in the following year. In the second section, wheat was cultivated in the winter season and in the summer season cowpea was intercropped with maize. This system proved to increase maize yield and reduced associated weeds (Zohry 2005). Maize and cowpea planting density under this system are 100 and 50% of its recommended density. In the third section, two mono crops were cultivated in the winter season, namely faba bean and clover. Maize was cultivated as a summer crop and short-season clover was cultivated as early winter crop after maize (Fig. 2).

Consequently, their results indicated that wheat yield was increased in the second rotation by 4% and maize yield was increase by 3%, and cotton yield was also increased by 4%, compared to its counterpart values under the prevailing crop rotation. Furthermore, soil organic matter content was increased by 22%, compared to the prevailing crop rotation and the value of soil available N, P, and K were increased

Year 1	Year 2	Year 3
Cotton relay intercropped on onion then sesame intercropped with cotton	Faba bean Maize Short-season clover	Wheat Cowpea intercropped with maize
Wheat Cowpea intercropped with maize	Cotton relay intercropped on onion then sesame intercropped with cotton	Faba bean Maize Short-season clover
Faba bean Maize Short-season clover	Wheat Cowpea intercropped with maize	Cotton relay intercropped on onion then sesame intercropped with cotton

Fig. 2 The second cotton crop rotation

Year 1	Year 2	Year 3
Cotton relay intercropped on wheat then cowpea intercropped with cotton	Faba bean intercropped with onion Cowpea intercropped with maize	Faba bean intercropped with wheat Soybean intercropped with maize Short-season clover
Faba bean intercropped with wheat Soybean intercropped with maize Short-season clover	Cotton relay intercropped on wheat then cowpea intercropped with cotton	Faba bean intercropped with onion Cowpea intercropped with maize
Faba bean intercropped with onion Cowpea intercropped with maize	Faba bean intercropped with wheat Soybean intercropped with maize Short-season clover	Cotton relay intercropped on wheat then cowpea intercropped with cotton

Fig. 3 The third cotton crop rotation

by 41, 39 and 31%, respectively, compared to the prevailing crop rotation. Additionally, in the second crop rotation, the dry weight of the annual weeds associated with wheat, maize, and cotton were significantly reduced by 35, 53 and 57%, respectively, compared to the prevailing crop rotation.

The intensification index was 2.50, which implied an increase in land productivity under this rotation, compared to the prevailing rotation. In addition, the total cereal units for this rotation was increased from 233.99 for the prevailing rotation to 279.8 for the second rotation, or by 20%, compared to its value under the prevailing rotation. Similarly, the farmer's net revenue obtained from this rotation was increased by 20%, compared to its value under the prevailing rotation.

The third crop rotation was a highly intensive rotation (contained intercropping systems and three-crop sequence, mainly legumes) and the crops were cultivated on raised beds (Fig. 3). In the first section, cotton was relay intercropped with onion (Zohry 2005), then cowpea was intercropped with cotton after the harvest of wheat.

In the second section (Fig. 3), faba bean was intercropped with wheat in the winter season (Abdel-Wahab and El Manzlawy 2016a, b). In this system, 100% of wheat recommended planting density was cultivated, where six rows above the raised beds were planted. Faba bean planting density under this system was 50% of its recommended density. Faba bean was planted on both sides of the raised beds. In the summer season, soybean was intercropped with maize. In this system, soybean was cultivated in the first week of May on the top of the raised beds in two rows and maize was cultivated on the both edges of the raised beds, 21 days after soybean planting. Planting densities of soybean and maize were 50% of its recommended densities (Sherif and Gendy 2012). Additionally, short-season clover was cultivated as an early winter crop. In the third section, faba bean was intercropped with onion in the winter season (Abou-Keriasha et al. 2013), where 75% of onion recommended planting density was cultivated, in four rows above the raised beds. Faba bean planting density was 50% of its recommended density planted on both sides of the raised beds. Furthermore, in the summer season, cowpea was intercropped with maize (Fig. 3).

Their results showed that wheat, maize and cotton yields were increased by 1, 3 and 8%, compared to its counterpart value under the prevailing crop rotation. Furthermore, soil organic matter content was increased by 38%, compared to the prevailing crop rotation. In addition, available N, P, and K was increase by 43, 45, and 35%, respectively compared to the prevailing crop rotation. These results were attributed to the increase in the number of legume crops in the rotation. In addition, the dry weight of annual weeds associated with wheat, maize and cotton were significantly reduced by 51, 70 and 66%, respectively, compared to the prevailing crop rotation. The cultivation of three-crop sequence including legumes and cotton and maize intercropping systems are the reasons for low percentage of the associated weeds, where year-long soil cover was provided suppressed associated weeds for the cultivated crops.

Additionally, their results also indicated that the intensification index of this highly intensive crop rotation was 2.64%, which implied high increase in land productivity. Moreover, the total value of cereal units was 289.7, which increased by 24%, compared to its value under the prevailing rotation. The farmer's net revenue of this rotation was increased by 32%, compared to its value under the prevailing rotation.

8 Calculation of the Applied Irrigation Water and Water Use Efficiency of These Three Rotations

For these three crop rotations, we calculated the applied irrigation water using BISm model (Snyder et al. 2004). The weather data from 2006 to 2008 were used in calculation of evapotranspiration and the applied water to each crop was calculated (Table 3). The BISm model calculated evapotranspiration (ETo) using Pennman-Montieth equation (Allen et al. 1998). It uses planting and harvest dates of each crop to determine the values of crop coefficients (Kc) and then determines water consumptive use. Water use efficiency was calculated by dividing total cereal units of each rotation by its applied irrigation water.

Table 3 The applied irrigation water to the cultivated crops in the first rotation

Crop	Water requirements (m^3/ha)
Clover (one cut)	4,071
Cotton	18,400
Wheat	6,171
Maize	10,686
Faba bean	4,657
Maize	10,686
Total	54,671

Table 4 The applied irrigation water (AW, m^3/ha) to the cultivated crops in the second and third crop rotations

Second rotation	AW	Third rotation	AW
Cotton relay intercropped on onion	19,391	Cotton relay intercropped on wheat	17,596
Sesame intercropped with cotton	14,720	Cowpea intercropped with cotton	14,720
Wheat	4,937	Faba bean intercropped with wheat	4,937
Cowpea intercropped with maize	8,549	Soybean intercropped with maize	8,549
Faba bean	3,726	Short-season clover	2,052
Maize	8,549	Faba bean intercropped with onion	6,731
Short-season clover	2,052	Cowpea intercropped with maize	8,549
Total	61,923		63,134

For the prevailing crop rotation, it contained six cultivated crops. The total applied irrigation water for this crop rotation was 54,671 m^3/ha. Furthermore, water use efficiency was 4.28 CU/mm.

In the second crop rotation, 5 crops were included within the three intercropping systems and 4 monoculture crops. The total applied water was increased by 13%, compared to the first rotation, with a total applied irrigation water equal 61,923 m^3/ha (Table 4). Whereas, the total number of crops in the third crop rotation was 12 crops and its applied water was increased by 15%, compared to the first rotation, with a total applied irrigation water equal to 63,134 m^3/ha (Table 4). However, water use efficiency for the second and the third rotations was increased to be 4.52 and 4.59 CU/mm, which implied that highly intensive crop rotation is the most advantageous type of crop rotations that increase farmer's net revenue and water use efficiency.

8.1 Cereal Crops Rotations

Two-year and three-year cereal crops rotations were implemented by Abou-Kersha et al. (1998). The two-year crop rotation contained wheat cultivated once and maize cultivated twice. The effect of preceding crops on maize yield was examined in the rotation. Wheat and onion preceded maize in the first section of the rotation and in the second section of the rotation, faba bean and clover preceded maize. Their results showed that the yield of maize was the highest after faba bean, followed by clover, where maize yield was increased by 41 and 26% after faba bean and clover, respectively compared to its cultivation after wheat.

In the three-year cereal crop rotation, wheat was cultivated once and maize was cultivated three times. Wheat preceded maize and soybean in the first section. Onion and faba bean preceded maize and soybean in the second section. Faba bean and clover preceded maize and soybean in the third section. Their results showed that thc yield of maize was the highest after faba bean, followed by clover. Maize yield was increased by 41 and 26% after faba bean and clover, respectively compared to its cultivation after wheat.

A comparison between rice two-year crop rotation with its counterpart three-year crop rotation revealed that rice yield was increased by 29% under the three-year crop rotation, compared to its yield under the two-year crop rotation, which was higher than rice yield produced under mono cultivation (Kamel et al. 2010).

8.2 Legume Crops Rotation

In experiments conducted by Abd El-Hadi et al. (2000), two cycles of three-year crop rotations (prevailing and proposed) were implemented to improve soil fertility by inclusion of legume crops in an intensive cropping systems. The prevailing crop rotation included clover (2 cuts) followed by cotton, wheat preceded with maize, where both are exhausting crops to the soil and faba bean preceded tomato. The proposed legume crops rotation, contained soybean, cowpea and faba bean. The values of N, P and K were increased by 18, 32 and 24% in the first cycle, respectively compared to the prevailing rotation under the first cycle. Furthermore, after the second cycle, N, P and K values were increased by 15, 47 and 18%, respectively, compared to the prevailing rotation. Furthermore, soil organic matter were increased by 17% after the second cycle, compared to the first cycle.

8.3 Crop Rotations to Defeat Diseases

It was reported that the deterioration of soil microbial community (Wang and Chen 2005) and reduction in the diversity of bacteria in both species and quantity, lower the number of fungi species and increase mould quantity (Xie et al. 2007) are the reasons for the decline in peanut yield after its continuous cultivation for several years. Zohry (2017) implemented a peanut three-year crop rotation to increase its yield after high incidence of nematode infection, where continuous cultivation of peanut results in continuous decline in the yield from 4.6 ton/ha in the first year of the cultivation to 0.82 ton/ha in the third year of cultivation, with 400% reduction. The implemented three-year crop rotation included non-host nematode crops to control nematode population in the soil, namely maize, sesame, and wheat. In addition, peanut was cultivated every three years. The results indicated that after implementing the suggested crop rotation, nematode number was reduced from 8800 nematode/250 g soil to 360 nematode/250 g soil. Consequently, peanut yield was increased from 0.82 ton/ha to 2.2 ton/ha.

Similarly, another crop rotation was implemented for sugar beet to reduce the incidence of nematode infection, as a result of sugar beet continuous cultivation every year. In the prevailing crop rotation, nematode population were tremendously increased to 9600 nematode/250 g soil and low sugar beet yield was produced with low quality, namely 24 ton/ha. In the implement crop rotation, sugar beet was cultivated once every three with non-host nematode crops. The nematode population was reduced to 710 nematode/250 g soil. Furthermore, higher sugar beet yield was obtained, namely 43 ton per hectare, with 80% increase and the quality of tubers were also improved (Zohry 2017).

8.4 Crop Rotations to Conserve Irrigation Water

El-Masry et al. (2010) implemented a three-year crop rotation in saline-sodic soil (EC = 6.1 dS/m, pH = 8.9 and ESP = 21.1%) to reduce the applied irrigation water to the prevailing rotation by replace high-water consuming crops with low-water consuming crops and in the main time reduce the salinity level by application of soil amendments. The prevailing rotation contained both paddy rice and clover both cultivated twice, which resulted in the application of high total applied irrigation water.

In the proposed crop rotation, clover was cultivated once and the second one was replaced by sugar beet. The same was done for paddy rice, which was cultivated once and the second one was replaced by sunflower.

Their results indicated that the applied irrigation water to the proposed rotation was lower than the applied amounts to the prevailing rotation. Furthermore, the application of gypsum and manure as soil amendment improved the soil of the proposed rotation after its end in the fourth year, where soil EC, pH and ESP became 4.1 dS/m, 7.8 and 12.6%, respectively.

Kamel et al. (2010) compared between the prevailing two-year paddy rice crop rotation in saline-sodic soil, where long season paddy rice cultivar was cultivated twice and between a proposed two-year rice crop rotation, where short season paddy rice cultivar was cultivated once. Furthermore, they also used soil amendments, namely gypsum and manure to overcome soil salinity. Their results indicated that cultivation of short season paddy rice once instead of cultivating long season paddy rice twice in the proposed crop rotation contributed in saving 19% of the applied irrigation water.

Similar procedures were done by Kamel et al. (2010) for the prevailing three-year paddy rice rotation to form a proposed three-year paddy rice rotation. In the proposed rotation, clover was cultivated once instead of three times as in the prevailing rotation, and was replaced by sugar beet. Paddy rice was cultivated once instead of three times and was replaced by cotton and sunflower. In addition, short season paddy rice cultivar replaced the long season cultivar in the prevailing rotation. Thus, the total saving in the applied irrigation water reached 29%.

9 Conclusions

To attain the first pillar of climate-smart agriculture, namely sustainably increase agricultural production, crop-smart, water-smart and soil-smart practices should be integrated. Several agricultural practices can realize this integration. Implementing intercropping systems accomplish yield advantage, water saving, increase soil fertility and reduce erosion, and reduce infestation of crops diseases. Furthermore, choosing the appropriate sequence of the cultivated crops contributes in increasing lands productivity and improving soil fertility. Crop rotations also can play an important role in increasing crops production through inclusion of crop-smart, water-smart and soil-smart practices.

It was reported that the implemented crop rotations in Egypt achieved yield increase, soil fertility improvements, reduction in diseases, and saving in the applied irrigation water.

References

Abbady K, El-Maaz E, Ahmed H, Zohry A (2016) Carbon sequestration as a function of intercropping management practice and different nitrogenous fertilizer types. J Soil Sci Agric Eng 7(8):565–586

Abd El-Hadi AH, Abo El-Enein RA, Omran A (2000) Sustainability of soil fertility status after three-year crop rotation at Middle Egypt region. In: Xth International colloquium for the optimization of plant nutrition, April 8–13, Cairo, Egypt

Abd El-Hadi AH, Abd El-Shafy AM, Aboelenin R, Haddad N, Salh A (2002) Soil fertility status as affected by different fertilizer levels and five crop rotations under rain fed conditions at North Sinai. Egypt J Soil 42(3):463–477

Abd El-Zaher ShR, Shams AS, Mergheny MM (2013) Effect of intercropping pattern and nitrogen fertilization on intercropping wheat with tomato. Egypt J Appl Sci 28(9):474–489

Abdel-Wahab Sh, El Manzlawy A (2016a) Yield and quality of intercropped sunflower with soybean under different sunflower plant spacing and slow-release nitrogen fertilizer rates in sandy soil. Int J Appl Sci 2(3):32–43

Abdel-Wahab T, El Manzlawy A (2016b) Yield and quality of intercropped wheat with faba bean under different wheat plant densities and slow—release nitrogen fertilizer rates in sandy soil. Am J Exp Agric 11(6):1–22

Abdel-Wahab Sh, El-Sayed W, El Manzlawy A (2016) Influences of some preceding winter crops and nitrogen fertilizer rates on yield and quality of intercropped maize with cowpea. Am J Exp Agric 11(6):1–19

Abo-Khadra SH, Sorour SGR, Toaima SEA, El-Sherief DEE (2018) Influence of conventional and intensive cropping sequences on wheat productivity under different mineral nitrogen fertilizer rates. Alex J Agric Sci 63(6):373–379

Abou-Keriasha MA, Eisa NMA, El-Wakil NMH (2013) Effects of intercropping faba bean on onion and wheat with or without inoculated bacteria on yields of the three Crops. Egypt J Agron 35(2):169–182

Abou-Keriasha MA, Ibrahim ST, Mohamadain EEA (2011) Effect of cowpea intercropping date in maize and sorghum fields on productivity and infestation weed. Egypt J Agron 33(1):35–49

Abou-Keriasha MA, Mohamed WMA, Eiasa NM, Kamel AS (2012) Intensive crop rotations to improve agricultural production in Middle Egypt. Egypt J Agric Res 90(4):427–443

Abou-Kersha MA (1998) Contibution of intercropping in crop rotation to cotton traits and yield. J Agric Sci 23(9):3591–3600
Abou-Kersha MA, Zohry AA, Haikle MA (1998) Maize and soybean yield as affected by preceding crops and crop rotation. J Agric Sci 23(11):4721–4728
Allen RG, Pereira LS, Raes D, Smith M (1998) Crop evapotranspiration: Guideline for computing crop water requirements. FAO No56
Altieri MA (1999) The ecological role of biodiversity in agro-ecosystems. Agric Ecosyst Environ 74:19–31
Andersen MK (2005) Competition and complementarily in annual intercrops—the role of plant available nutrients. PhD thesis, Department of Soil Science, Royal Veterinary and Agricultural University, Copenhagen, Denmark. SamfundslitteraurGrafik, Frederiksberg, Copenhagen
Anderson RL (2007) Managing weeds with a dualistic approach of prevention and control: a review. Agron Sustain Dev 27:13–18
Batista TMV, Bezerra FN, Silva IN, Da Silva ML, De Oliveira EQ, Junior APB (2016) Agronomic efficiency of the intercropping of arugula with carrot under different population combinations. Rev Caatinga Mossoró 29(1):76–84
Bennett AJ, Bending GD, Chandler D, Hilton S, Mills P (2012) Meeting the demand for crop production: the challenge of yield decline in crops grown in short rotations. Biol Rev 87:52–71
Berzsenyi Z, Gyorffy B, Lap D (2000) Effect of crop rotation and fertilisation on maize and wheat yields and yield stability in a long-term experiment. Eur J Agron 13:225–244. https://doi.org/10.1016/S1161-0301(00)00076-9
Bruns HA (2012) Concepts in crop rotations, agricultural science. In: Godwin Aflakpui (ed). In Tech. ISBN: 978-953-51-0567-1. http://www.intechopen.com/books/agricultural-science/conceptsin-crop-rotation
Brankatschk G, Finkbeiner M (2014) Application of the Cereal Unit in a new allocation procedure for agricultural life cycle assessments. J Clean Prod 73:72–79
Brockhaus J (1962) ABC der Landwirtschaft, Band (i). A-K. VEB, Brock HausVerlag, Leipzig, pp 488–489
Camarotto C, Dal-Ferro N, Piccoli I et al (2018) Conservation agriculture and cover crop practices to regulate water, carbon and nitrogen cycles in the low-lying venetian plain. CATENA 167:236–249
Chikte P, Thakare SM, Bhalkare SK (2008) Influence of various cotton-based intercropping systems on population dynamics of thrips, Scircothrips dorsalis Hood and whitefly. Bemisia Tabaci Genn Res Crop 9:683–687
Chimonyo VGP, Modi AT, Mabhaudhi T (2016) Assessment of sorghum-cowpea intercrop system under water-limited conditions using a decision support tool. Water SA 42(2):316–327. https://doi.org/10.4314/wsa.v42i2.15
Choudhary VK, Choudhury BU (2016) A staggered maize–legume intercrop arrangement influences yield, weed smothering and nutrient balance in the eastern Himalayan region of India. Exp Agric 54:181–200
Coll L, Cerrudo A, Rizzalli R, Monzon JP, Andrade FH (2012) Capture and use of water and radiation in summer intercrops in the south-east Pampas of Argentina. Field Crops Res 134:105–113
Dwivedi A, Dev I, Kumar V, Yadav RS, Yadav M, Gupta D, Singh A (2015) Potential role of maize-legume intercropping systems to improve soil fertility status under Smallholder farming systems for sustainable agriculture in India. Tomar1 Inter J Life Sci Biotechnol Pharma Res 4(3):215–222
El-Fakharanya SKM, Samy MA, Ahmed SA, Khattab MA (2012) Effect of intercropping of maize, bean, cabbage and toxicants on the population levels of some insect pests and associated predators in sugar beet plantations. J Basic Appl Zool 65:12–28. https://doi.org/10.1016/j.jobaz.2012.02.002
El-Masry MA, Kamel AS, Zohry AA (2010) Sustainable production of cotton in saline sodic soil in Northern Part of Nile Delta in Egypt. Egypt J Argon 32(1):59–72
El-Mehy AA, Mohamed MHM (2018) Yield and economic evaluation of maize and tomato as affected by cropping systems and some growth stimulants. Middle East J Appl Sci 08(01):209–222

El-Mehy A, Abd Allah AM, Mohamed TS, Kasem EE (2020) Intercropping of some faba bean cultivars with sugar beet using different irrigation intervals under sprinkler system in sandy soils of plant production. J Plant Prod 11(12):1215–1225

FAO (Food and Agriculture Organization of the United Nation) (2003) On-farm trials for adapting and adopting good agricultural practices. Food and Agriculture Organization of the United Nations, Rome

Farghly BS, Zohry AA (2002) Effect of preceding crops in three-year crop rotations and N rates on yield and yield components of Egyptian cotton. Egypt J Agric Res 80(1):293–307

Feng L, Sun Z, Zheng M, Muchoki M, Zheng J, Yang N, Bai W, Feng CH, Zhang Z, Cai Q, Zhang D (2016) Productivity enhancement and water use efficiency of peanut-millet intercropping. Pak J Bot 48(4):1459–1466

Francis CA (1986) Future perspectives of multiple cropping. In: Francis CA (ed) Multiple cropping systems. Macmillan, New York, pp 351–370

Gan Y, Hamel C, O'Donovan JT, Cutforth H, Zentner RP, Campbell CA, Niu Y, Poppy L (2015) Diversifying crop rotations with pulses enhances system productivity. Sci Rep 5, Article number: 14625. https://doi.org/10.1038/srep14625

Gitari HI, Gachene CKK, Karanja NN, Kamau S, Nyawade S, Sharma K, Schulte-Geldermann E (2018) Optimizing yield and economic returns of rain-fed potato (Solanum tuberosum L.) through water conservation under potato-legume intercropping systems. Agric Water Manag 208:59–66

Gonzalez-Dıaz L, van den Berg F, van den Bosch F, Gonzalez-Andujar JL (2012) Controlling annual weeds in cereals by deploying crop rotation at the landscape scale: Avena sterilis as an example. Ecol Appl 22:982–992

Hamd-Alla WA, Shalaby EM, Dawood RA, Zohry AA (2014) Effect of cowpea (Vigan sinensis L.) with maize (Zea mays L.) intercropping on yield and its components. Inter Scho Sci Res Innov 8(11):1170–1176

Hamdany MKH, Aassar MR (2017) Effect of intercropping three faba bean varieties with sugar beet plants on piercing sucking insect pests and associated natural enemies under ridge space and seedling rates in relation crop yield. Egypt Acad J Bio Sci 10(6):57–77

Hauggaard-Nielsen H, Mundus S, Jensen ES (2009) Nitrogen dynamics following grain legumes and subsequent catch crops and the effects on succeeding cereal crops. Nutr Cycl Agroecosyst 84(3):281–291. https://doi.org/10.1007/s10705-008-9242-7

Huang Q, Yin Z, Tian C (2001) Effect of two different straw mulching methods on soil solute salt concentration. Arid Land Geogr 24:52–56

Ibrahim S, Shaaban M, Gendy EK (2010) Intercropping faba bean with tomato. Egypt J Appl Sci 25(6A):167–181

Iqbal N, Hussain S, Ahmed Z, Yang F, Wang X, Liu W, Yong T, Du J, Shu K, Yang W (2018) Comparative analysis of maize–soybean strip intercropping systems: a review. Plant Prod Sci 22:131–142

Jaya IKD, Bell CJ, Sale PW (2001) Modification of within-microclimate in maize for intercropping in the lowland tropics. In: Proceedings of the Australian agronomy conference. Australia Society of Agronomy, Australia, pp 123–133

Kamel AS, Sahar AS, Ibrahim ST (2004) New agro-techniques in intensive crop rotations under marginal conditions of Upper Egypt. New directions for a diverse planet: Proceedings of the 4th international crop science congress, Brisbane, Australia, 26 Sep–1 Oct 2004

Kamel AS, El-Masry ME, Khalil HE (2010) Productive sustainable rice based rotations in saline-sodic soils in Egypt. Egypt J Agron 32(1):73–88

Kariaga BM (2004) Intercropping maize with cowpeas and beans for soil and water management in Western Kenya. In: Proceedings of the 13th international soil conservation organization conference, Conserving Soil and Water for Society, Brisbane, July, pp 1–5

Katyayan A (2005) Fundamentals of agriculture. Kushal Publications & Distributors, Varanasi, Uttar Pradesh, pp 10–11

Kaye NM, Mason SC, Jackson DS, Galusha TD (2007) Crop rotation and soil amendments alters sorghum grain quality. Crop Sci 47:722–729

Kestha MM, El-Baz MG (2004) Studies on sunflower-tomato intercropping. In: Proceedings of the 16th international sunflower conference, Fargo, ND USA, pp 514–518

Khalifa HE, Sherif MA, El-Melegy A, Abo Elenien RA, Heliela H, Owies T (2004) Crop rotations and fertilizer type effects on barley production and water-use efficiency at the rain fed area of North Sinai. In: International workshop on management of poor quality water for irrigation; institutional, health and environmental aspects, Moscow, Russia, pp 241–249

Kiboi MN, Ngetich KF, Diels J et al (2017) Minimum tillage, tied ridging and mulching for better maize yield and yield stability in the central highlands of Kenya. Soil Tillage Res 170:157–166

Kirkegaard JA, Christen O, Krupinsky J, Layzell D (2008) Break crop benefits in temperate wheat production. Field Crops Res 107:185–195

Kumpawat BS, Rathore SS (2003) Effect of preceding grain legumes on growths, yield, nutrient content and uptake by wheat between different nitrogen levels. Field Crop Res 25(2):209–214

Lamlom M, Abdel-Wahab Sh, Abdel-Wahab T, Gendy E (2015) Residual effects of some preceded winter field crops on productivity of intercropped soybean with three maize cultivars. Am J BioSci 3(6):226–242. https://doi.org/10.11648/j.ajbio.20150306.15

Lithourgidis AS (2011) Annual intercrops: an alternative pathway for sustainable agriculture. Aust J Crop Sci 5(4):396–410

Machado S (2009) Does intercropping have a role in modern agriculture? J Soil Water Conserv 64(2):233–239

Macák M, Žák S, Andrejčíková M (2015) Productivity and macro elements content of cereal and legume crops. Acta Fytotechn Zootechn 18:160–162

Maitra S (2020) Intercropping of small millets for agricultural sustainability in dry lands: a review. Crop Res 55:162–171

Malik R (2010) Soil quality benefits of break crops and/or crop rotations: a review. In: 19th World congress of soil science, soil solutions for a changing world, 1–6 August, Brisbane, Australia

Mao LL, Zhang LZ, Li WW, Werf WVD, Sun JH, Spiertz H, Li L (2012) Yield advantage and water saving in maize/pea intercrop. Field Crop Res 138:11–20

Mead R, Willey RW (1980) The concept of a "land equivalent ratio" and advantages in yields from intercropping. Exper Agric 16:217–228

Megawer EA, Sharaan AN, El-Sherif AM (2010) Effect of intercropping patterns on yield and its components of barley, lupine or chickpea grown in newly reclaimed soil. Egypt J Appl Sci 25(9):437–452

Miao Q, Rosa R, Shi H, Paredes P, Zhu L, Dai J, Gonçalves JM, Pereira LS (2016) Modeling water use, transpiration, and soil evaporation of spring wheat–maize and spring wheat–sunflower relay intercropping using the dual crop coefficient approach. Agric Water Manag 165:211–229

Morgado LN, Carvalho CF, Souza B, Santana MP (2002) Fauna of bees (Hymenoptera: Apoidea) on sunflower flowers, Helianthus annuus L., in Lavras-MG. Ciência Agrotecnologia 26:1167–1177

Nasri R, Kashani A, Barary M, Paknejad F, Vazan S (2014) Nitrogen uptake and utilization efficiency and the productivity of wheat in double cropping system under different rates of nitrogen. Int J Biosci 4:184–193

Nicholls CI, Altieri MA (2013) Plant biodiversity enhances bees and other insect pollinators in agroecosystems. A review. Agron Sustain Dev 33:257–274

Nyawade SO, Gachene CK, Karanja NN, Gitari HI, Schulte-Geldermann E, Parker ML (2019) Controlling soil erosion in smallholder potato farming systems using legume intercrops. Geoderma Reg 17:e00225

Ouda S, Zohry A (2018) Cropping pattern to face water scarcity. In: Cropping pattern to overcome abiotic stresses: water, salinity and climate. Springer Publishing House. ISBN 978-3-319-69879-3

Ouda S, Hefny YAA, Abdel-Wahab TI, Abdel-Wahab SI (2018) Intercropping systems of sunflower and peanut under different irrigation regimes and potassium fertilizer levels. Egypt J Agron Crop Sci 85–104

Ouda S, Zohry A, Abdel-Wahab S (2020) Sustainable intensive cropping to reduce irrigation-induced erosion: changing cropping sequence under sprinkler irrigation practice. Moroccan J Agric Sci 1(1):26–33

Ouma G, Jeruto P (2010) Sustainable horticultural crop production through intercropping: the case of fruit and vegetable crops: a review. Agric Biol J North Am 1:1098–1105

Parthiban P, Chinniah C, Murali Baskaran RK, Suresh K, Kumar AR (2016) Impact of intercropping system to minimize the sucking pests incidence in groundnut (Arachis hypogaea Linnaeus). Legume Res LR-3770:1–4

Porto VCN, Bezerra NF, Lima JSS, Barros JAP, Moreira JN (2011) Combination of lettuce and rocket cultivars in two cultures intercropped with carrots. Horti Bras 29:404–411

Qin A, Huang G, Chai Q, Yu A, Huang P (2013) Grain yield and soil respiratory response to intercropping systems on arid land. Field Crops Res 144:1–10. https://doi.org/10.1016/j.fcr.2012.12.005

Selim M, Shams AS (2019) Maximizing efficiency of land and water utilization and profitability of interplanting maize with mandarin trees using irrigation with fish waste water under sandy soil and drip irrigation conditions. Middle East J Agric Res 08(04):1240–1252

Shams AS, Kamel AS (2014) Rotations in Coastal plains to combat desertification in Egypt. Int J Water Resour Arid Environ 3(2):121–131

Shams AS, El-Debaby AS, Roshdy A, Kamel AS (2012) Effect of maize plants distribution and nitrogen fertilization levels in peanut-maize intercrop. Egypt J Agron 34(1):39–52

Sheha AM, Ahmed NR, Abou-Elela AM (2014) Effect of crop sequence and nitrogen levels on rice productivity. Ann Agric Sci 52(4):451–460

Sherif SA, Gendy EK (2012) Growing maize intercropped with soybean on beds. Egypt J Appl Sci 27(9):409–423

Silva PSLI, Oliveira OFII, Silva PIBIII, Silva KMBIV, Braga JDV (2009) Effect of cowpea intercropping on weed control and corn yield. Planta Daninha 27(3):349–358

Singh BB, Ajeigbe HA, Tarawali SA, Fernandez-Rivera S, Abubakar M (2003) Improving the production and utilization of cowpea as food and fodder. Field Crop Res 84:169–177

Squire GR, Rodgers S, Wright G (2000) Community-scale seed bank response to less intense rotation and reduced herbicide input at three sites. Ann Appl Biol 136:47–57

Snyder RL, Orang M, Bali K, Eching S (2004) Basic irrigation scheduling (BISm). http://www.waterplan.water.ca.gov/landwateruse/wateruse/Ag/CUP/Californi/Climate_Data_010804.xls

Stone LF, Silveira PM (2001) Effects of soil tillage systems and crop rotations on soil porosity and bulk density. Rev Bras Ciênc Solo 25(2):345–351

Tanaka DL, Anderson RL, Rao SC (2005) Crop sequencing to improve use of precipitation and synergize crop growth. Agron J 97:385–390

Tang X, Zhang Y, Jiang J, Meng X, Huang Z, Wu H, He L, Xiong F, Liu J, Zhong R, Han Z, Tang R (2021) Sugarcane/peanut intercropping system improves physicochemical properties by changing N and P cycling and organic matter turnover in root zone soil. Peer J 9:e10880. https://doi.org/10.7717/peerj.10880

Thenail C, Joannon A, Capitaine M, Souchere V, Mignolet C, Schermann N, Di Pietro F, Pons Y, Gaucherel C, Viaud V, Baudry J (2009) The contribution of crop-rotation organization in farms to crop-mosaic patterning at local landscape scales. Agric Ecosyst Environ 131:207–219

Toaima SEA (2007) Effect of crop rotations and nitrogen fertilizer on yield, yield components and associated weeds in cotton. Arab Univ J Agric Sci 15(2):349–359

Tsubo M, Walker S (2002) A model of radiation interception and use by a maize–bean intercrop canopy. Agric Forest Meteorol 200:203–215

Varvel GE (2000) Crop rotation and nitrogen effects on normalized grain yields in a long-term study. Agron J 92:938–941

Wang MZ, Chen XN (2005) Obstacle and countermeasure of sustainable high yield for peanut in low-hilly red soil region. J Peanut Sci 34(2):17–22

Wolka K, Mulder J, Biazin B (2018) Effects of soil and water conservation techniques on crop yield, runoff and soil loss in Sub-Saharan Africa: a review. Agric Water Manag 207:67–79

Xie H, Wang XX, Dai CC, Chen JX, Zhang TL (2007) Effects of peanut (Archishypogaea) intercropped with medicinal plants on soil microbial community (briefing). Chin J Appl Ecol 18(3):693–696

Yang C, Huang G, Chai Q, Luo Z (2011) Water use and yield of wheat/maize intercropping under alternate irrigation in the oasis field of northwest China. Field Crops Res 124:426–432

Zhang X, Chen P, Du Q, Zhou Y, Ren J, Jin F, Yang W, Yong T (2019) Effects of maize/soybean and maize/peanut intercropping systems on crops nitrogen uptake and nodulation nitrogen fixation. Chin J Eco-Agric 27:1183–1194. https://doi.org/10.13930/j.cnki.cjea.181055

Zhou Q, Chen J, Xing Y, Xie X, Wang L (2019) Influence of intercropping Chinese milk vetch on the soil microbial community in rhizosphere of rape. Plant Soil 440:85–96. https://doi.org/10.1007/s11104-019-04040-x

Zohry AA (2005) Effect of relaying cotton on some crops under bio-mineral N fertilization rates on yield and yield components. Ann Agric Sci 431:89–103

Zohry A, Ouda S (2019) Intercropping systems for sugar beet to improve its land and water productivity. J Soils Crops 29(2):218–226

Zohry AA (2017) Crop Intensification and crop rotation: unconventional methods for agricultural development. (التكثيف المحصولى و الدوره الزراعيه: طرق غير تقليديه للتنميه الزراعيه). Noor Publishing. ISBN 978-3-330-79924-0 (in Arabic)

Zohry AA, Ouda S, Hamd-Alla W, Shalaby E (2017a) Evaluation of different crop sequences for wheat and maize in sandy soil. Acta Agric Slov 109(2):383–392

Zohry AA, Abbady K, El-Maaz E, Ahmed H (2017b) Maximizing land productivity by diversified cropping systems with different nitrogen fertilizer types. Acta Agric Slov 109(3):482–492

Zohry A, Ouda S, Abdel-Wahab T (2020a) Sustainable intensive cropping to reduce irrigation-induced erosion: Intercropping systems under surface irrigation practice. Moroccan J Agric Sci 1(2):63–71

Zohry A, Hefny Y, Ouda S (2020b) Interplanting four legume crops under orange trees using different irrigation water and fertilizer sources in sandy soil. In: 16th International conference on crop science, 10th of October, Agron. Dept., Fac. Agric, Cairo, Al-Azhar Univ., Oct. 13, 2020, Cairo, Egypt, pp 341–346

Part II
Build Resilience to Climate

Chapter 5
Climate Extremes and Crops

Samiha Ouda and Abd El-Hafeez Zohry

Abstract Climate resiliency is the capacity for a socio-ecological system to absorb stresses and maintain function in the face of external stresses imposed on it by climate change, and adapt, reorganize and evolve into more desirable configurations that improve the sustainability of the system, leaving it better prepared for future climate change impacts. In this chapter, we discussed the implication of climate variability and extreme weather events on water resources and crops production, with emphasis on Egyptian conditions. The concept of building resilience to climate variability stress were discussed in the context of reducing exposure, reducing sensitivity and increasing adaptive capacity. Reducing exposure can be achieved by implementing resilient farming systems based on increased levels of biodiversity. Reducing sensitivity could occurred by selection of tolerant crops and varieties, and increasing adaptive capacity of crops could accomplished by early projections of extreme weather events. Examples implemented in Egypt were discussed.

Keywords Climate variability · Extreme weather events · Resilient farming system · Resilient crops · Early warning system

1 Introduction

Weather is the main factor controls crops production. Crop development and season length is dependent on weather. IPCC (2007) stated that "the manner in which human beings respond to climate variability is critical not only to survival but also to well-being". Climate variability is often used to describe any kind of change in climate that may be natural or human-induced (Hegerl et al. 2000). The term "climate variability" is often used to denote deviations of climate statistics over a given short period of

S. Ouda (✉)
Water Requirements and Field Irrigation Research Department, Soils, Water and Environment Research Institute; Agricultural Research Center, Giza, Egypt

A. E.-H. Zohry
Crops Intesefication Research Department, Field Crops Research Institute; Agricultural Research Center, Giza, Egypt

S. Ouda and A. E.-H. Zohry, *Climate-Smart Agriculture*,
https://doi.org/10.1007/978-3-030-93111-7_5

time (such as a specific month, season, or year) from the long-term climate statistics relating to the corresponding calendar period (Sami Ul Din et al. 2019).

Extreme weather events, such as droughts, heat and cold waves, and flooding have serious effects on food production. IPCC (2011) defines the extreme weather events caused by climate variability "as rare as or rarer than the 10th or 90th percentile of a probability density function estimated from observations". It is expected that the extreme weather events will increase worldwide, therefore anticipating their effects on crop yields is important for food security.

Human suffering and poverty are resulted from climate-related risks (IPCC 2014). In developing countries, the increasing incidence and magnitude of extreme weather events stand as a potential disaster for food production (Sivakumar et al. 2011). Furthermore, it is expected that the frequency and magnitude of extreme weather events will be intensify under climate change (Solomon et al. 2007). The primary climate challenges that impact farming are seasonal shifts, ecological disturbances, and extreme weather events, which impact the principle needs of a crop namely water, sun and soil (Meuwissen et al. 2019). If agricultural ecosystems are disrupted, crop yields could considerably pull apart, which directly impacts food availability (Matchett et al. 2015). Shifts in seasonal weather can cause several problems, such as increased atmospheric carbon dioxide, increased daytime and nighttime temperatures, and variability in rainfall patterns (Meuwissen et al. 2019).

Climate resiliency was identified by Smith (2012) as "the capacity for a socio-ecological system to absorb stresses and maintain function in the face of external stresses imposed on it by climate change, and adapt, reorganize and evolve into more desirable configurations that improve the sustainability of the system, leaving it better prepared for future climate change impacts". Building resilience to climate is an important goal in climate-smart agriculture. Climate-smart agriculture is an approach or set of practices aimed at sustainably increasing agricultural productivity and incomes sustainably, while building resilience and adapting to climate variability as well as reducing and/or removing greenhouse gas emissions where possible. Building resilience of the cultivated crops is enhancing its ability to adapt to climatic shocks, which can be improved through internal and external interventions (Meuwissen et al. 2019). Developing resilient strategies for agriculture is highly dependent on knowledge of how and where the climate has varied. Three stages can be done to achieve resilience building, namely reducing exposure, reducing sensitivity, and increasing adaptive capacity (Gitz and Meybeck 2012). Adaptation is defined as the "decision-making process and the set of actions undertaken to maintain the capacity to deal with future changes or perturbations to socio-ecological system without undergoing significant changes in function, structural identity, while maintaining the option to develop" (Nelson et al. 2007).

In this chapter, the concept of building resilience to climate variability and extreme weather events were discussed in the context of reducing exposure by implementing resilient farming systems based on increased levels of biodiversity, reducing sensitivity by selection of tolerant crops and varieties, and increasing adaptive capacity of crops by early projections of extreme weather.

2 Climate Variability Leads to Extreme Weather Events

Variability in weather and climate inherently leads to the occurrence of extreme weather or climate events. The increasing frequency of extreme weather events is mainly related to climate change. These events, more unusual and more severe than normal or average weather, include heat waves (high-temperature events), cold waves (low-temperature events), downpours (heavy-precipitation events) and droughts (low-precipitation events) (van der Wiel and Bintanja 2021). Extreme weather events is negatively influence crop yields under both irrigated and rain fed agriculture (Adamgbe and Ujoh 2013).

It is expected that the intensity and frequency of extreme precipitation events will increase, particularly in winter months (Sugiyama et al. 2014). With episodes of extreme climate expected to become more frequent (Bathiany et al. 2018), measures to buffer crop production against these stresses are a critical aspect of climate adaptation (Davis et al. 2019). Therefore, anticipating and calculating their effects on crop yields is important for food security (Porter et al. 2014). Evaluation of climate-induced crop damages is imperative to formulate innovative technologies and management strategies to reduce the vulnerability of farms and agriculture (Lobell et al. 2013). Most crops are sensitive to direct effects of high temperature, decreased precipitation, flooding, and untimely freezes during critical growth phases (Motha 2011).

Moreover, insect, pest, and plant diseases may magnified under exposure to extreme weather events (Motha 2011). The outbreaks of two-spotted spider mites damaging soybeans throughout the entire Midwest region existed, where the damage occurred during the critical flowering, pod development, and pod filling growth stages (Rosenzweig et al. 2001). Other indirect effects of extreme weather events could occur, such as influences on soil processes, and nutrient dynamics.

3 Effects of Climate Variability on Crops

3.1 *Effects of Heat Waves*

Extreme events are expected to cause instability of crops yields and forming high risk to global crop production (Lobell et al. 2013). Crop yields show a strong correlation with temperature change and with the duration of heat waves, which differ based on plant maturity stages during extreme weather events (Hoffmann 2013). Schlenker and Roberts (2009) indicated that extremely high daytime temperatures are destructive and could be lethal to crops. Rosenzweig et al. (2001) reported that higher temperatures than threshold temperatures may lead to significant reductions in crop yields.

Due to differences in growth patterns among crops, the impact of warming temperatures and weather extremes is crop dependent. Physiological processes of a crop could be affected by extreme weather events and causing physical damage (van der Velde et al. 2012). Acceleration in phenological development of crops under warmer conditions is observed, which reduces growing season, reduce net photosynthesis and consequently reduce yield (Lobell and Ortiz-Monasterio 2007). Ben-Asher et al. (2008) indicated that photosynthesis rates of sweet corn was decreased by at least 30% when the temperature exceeded 30 °C. Furthermore, higher air temperatures than 36 °C cause pollen to lose viability in maize (Motha 2011).

High temperature also influences soil processes, nutrient dynamics, and pest organisms. Soil temperature higher than 20 °C depresses potato bulking (Wilhite et al. 2005).

3.2 *Effect of Precipitation Variability*

Low precipitation events have been attributed to falls in crop productivity (Sivakumar et al. 2005). Drought leads to stomata closure and photosynthesis reduction, which negatively affecting the yield (Chaves et al. 2010). Gornall et al. (2010) stated that even small changes in mean annual rainfall has an impact on productivity under rain fed agriculture. Yield losses higher than 30% in fruit and vegetable crops were found due to drought conditions in 2016 in the Northeastern United States was reported (Sweet et al. 2017). Soil moisture deficits are very detrimental to maize, where 4 days of soil moisture stress reduces yields up to 50% (Wilhite et al. 2005). Whereas, thunderstorms and hailstorms were found to cause a significant reduction in wheat yield in Pakistan even it was rated moderate or low in severity (Elahia et al. 2021).

3.3 *Interaction Between High Temperature and Precipitation Variability*

Temperature and rainfall affect the development of plants, either alone or by interacting with other factors (IPCC 2007). Severe changes in rainfall patterns coupled with rising temperatures result in unfavorable growing conditions and changes in the cropping calendar, which may subsequently reduce productivity (Manneh et al. 2010). The interaction between high temperature and drought stress reduced both wheat photosynthesis and yield (Shah and Paulsen 2003). Powell and Reinhard (2016) found that high temperature and precipitation events are time specific, where its effect on yields is varied based on growth stages of wheat. The severe drought of 1988 in the U.S. Midwest, accompanied by higher than normal temperatures, began early in the spring and continued throughout most of the summer affected agriculture and water resources and resulted in drops in crop yields by approximately 37%

(Wilhite et al. 2005). Flooding causes seedling death in maize and soybean; the combination of flooding with high temperature accelerates death.

4 Climate Variability and Agriculture in Egypt

Egyptian agriculture faces serious risks from climate variability causing significant negative impacts on fresh water resources, and on the yield of major crops, which consequently affect the situation of food security.

4.1 Climate Variability Effects on Fresh Water Resource

Climate variability affects water resource through its effects on rain patterns and variability in temperature levels. In Egypt, the Nile River represents the largest freshwater resource, with a minor contribution coming from rainfall and non-renewable groundwater aquifers. Extreme weather events, namely excessive rain fall and flash flood affects both surface water and ground water resources, with more damage to surface water. This situation creates many problems as the wastewater treatment plants are not designed to cope to these extreme weather events (Yehia et al. 2017). With excessive rainfall, runoff causes the transportation of solid material to surface water sources, which increase turbidity (Delpla et al. 2009).

Egypt has been subject to some flash floods, recently owing to extreme rainfall events, in particular regions, namely Sinai, North Coast, and Upper Egypt. Heavy rainfalls and flash floods are responsible for huge losses of infrastructure (Yehia et al. 2017). In regions characterized by very low annual precipitation, these episodes can be catastrophic in Egypt (Baldi 2019). The flash floods resulting from heavy rainfall has affected Egyptian coastal areas along the Mediterranean Sea and the Red Sea, as well as Upper Egypt (Al Zayed et al. 2013). It usually take place in autumn and winter, where extreme rainfall events cause flash floods over the Red Sea Mountains and the Sinai Peninsula that is highly variable in short duration (Moawad et al. 2016). The heavy rainfall across Egypt has caused rapid flooding affecting many regions of the country. This heavy rainfall swept away a great deal of dirt and dust from surrounding environments to the watercourses, which would have a great influence on the quality and quantity of the water (Yehia et al. 2017).

4.2 Climate Variability and Rain Fall in North Coast of Egypt

Higher variability of rain fall exists in the northern cost of Egypt due to the effect of climate change, as it shows from the available rain fall data in the period from 1997 to 2020. It is obvious from Fig. 1 that the annual rain fall value in north coast of Egypt

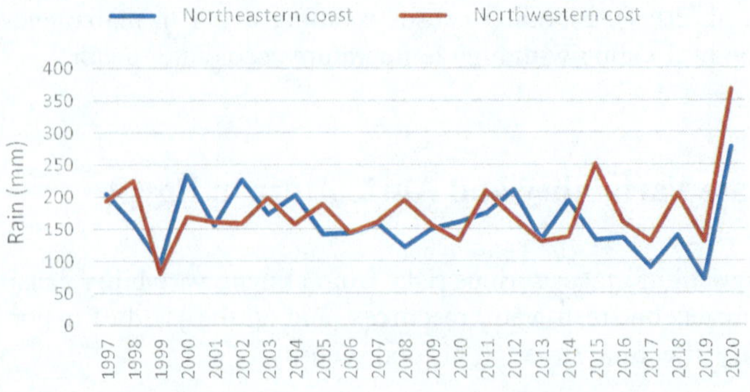

Fig. 1 Annual values of rain fall (mm) from 1997 to 2020 in northeastern and northwestern costs of Egypt

is highly variable from year to year. The highest rain fall values were found in 2020, namely 276 and 376 mm for northeastern and northwestern costs, respectively. The lowest values of in northeastern and northwestern coasts were found in 2019, namely 70 mm, with range between the highest and the lowest values equal to 206 mm.

In Northwestern cost, the lowest value of annual rain fall was found in 1999, namely 80 mm, with range between the highest and the lowest values equal to 287 mm (Fig. 1). The figure also showed that the level of dispersion around the mean was high presented by high value of coefficient of variation (CV%), namely 30 for and 31%.

The methodology of Kumar et al. (2009) was used to assess rain fall deviations from the mean. Four categories were assigned by the authors to monitor and evaluate the rain fall patterns as followed: (1) normal rain when ±10% deviation exists; (2) deficit rain when −10 to −60% deviation; (3) scanty rain when less than −60% exist; and (4) excess rain when deviation is greater than 20%.

Applying this procedure on the rain fall data in Egypt from 1997 to 2020 revealed that that rain fall in north cost of Egypt can be characterized by being deficit rain, where deviation from the mean were +17 and −21 in Northeastern and Northwestern costs, respectively. Thus, it is expected that rain variability in the north coast of Egypt will affect the productivity of the cultivated crops in these area, and the livelihood of the community.

5 Heat Waves and Crops Production in Egypt

5.1 Wheat Crop

Heat stress affects various wheat plant processes leading to morpho-physiological alterations in plants, hindering the development processes and eventually resulting in great yield loss (Grant et al. 2011). Almeselmani et al. (2009) observed that high temperature (35/25 °C) imposed after tillering showed a significant reduction in water potential in wheat, which reduces plant growth by promoting leaf senescence and abscission (Kosova et al. 2011). Heat stress in wheat during anthesis causes reduction in pollen tube development, and increases of pollen mortality (Oshino et al. 2011). Plants exposed to temperatures above 24 °C during reproductive stage significantly reduced grain yield and yield reduction continued with increasing duration of exposure to high temperature (Prasad and Djanaguiraman 2014).

In the winter growing season of wheat in 2010 in Egypt, temperature increased by 2.2 °C, compared to its value during winter 2010, which caused decrease in the national wheat production. This rise of temperature occurred during anthsis stage, which caused high yield reduction. In Upper Egypt, the highest reduction in wheat yield was observed, namely 21% and the lowest decrease by 8% were observed in Lower Egypt (Khalil and Hassanein 2016). Taha (2012) reported a reduction in wheat yield in north Egypt in 2010 by 12%, compared to the value obtained in 2009. Whereas, Saleh et al. (2017) assessed the averaged reduction in wheat production over Egypt in 2010 and stated that the average reduction was 15%.

5.2 *Rice Crop*

The most susceptible growth stage to heat stress in rice is flowering stage. The previous greenhouse experiments showed that temperatures of 32–36 °C cause high spikelet sterility (Satake and Yoshida 1978), which is associated with grain yield reduction. Mahmood et al. (2010) indicated that exposing rice plants to heat waves during flowering growth stage could cause sterility, whereas during grain filling stage it reduces grain weight and grain number, with a decline in grain number directly proportional to increasing temperature. High temperatures above 35 °C at the flowering stage inhibit anther dehiscence and causing incomplete fertilization (Jagadish et al. 2007).

It was also reported that in 1998 summer growing season in Egypt, rice plants were exposed to heat wave during flowering and grain filling stages caused reduction in national rice production (Khalil and Hassanein 2016).

5.3 *Maize Crop*

Maize yields decline with warmer temperatures due to acceleration of the crop's development, especially during the grain filling period (Motha 2011). Temperature above 35 °C for a prolonged period is considered unfavorable for maize crop growth and development and, particularly 40 °C during flowering and grain filling have severe negative impacts on grain yield (Shiferaw et al. 2011). Plants under heat stress exhibited significantly reduced stomatal conductance resulting in a reduced rate of photosynthesis. Excessive heat also causes a reduction in net photosynthesis, leaf area, reduced biomass accumulation and seed weight (Zhao et al. 2016). In Egypt, Saleh et al. (2017) indicated that maize national production was reduced by an average of 11% in 2015, as a result of heat wave occurred in that year, compared to its production in 2010.

6 Measures to Limit Climate Variability Stress in Egypt

Measures to limit the stresses resulted from climate variability have been extensively studied, including resilient farming systems based on increased levels of biodiversity (Altieri et al. 2015), and selection for new crops and tolerant varieties (Yadav et al. 2010). Furthermore, early projections of extreme weather events is very important action to be done to face climate variability (Tian et al. 2018).

In Egypt, several measures have been done on field level to reduce exposure and sensitivity of plants community to climate variability, as well as increasing adaptive capacity.

6.1 Implementing Resilient Farming System

The ability of farming systems to cope with challenges can be conceptualized as resilience that may be conferred by diversifying crops and livestock and/or diversifying crops and fish by farmers implementing adaptive approaches in response to stresses (Bullock et al. 2017). Increasing the farm's agricultural biodiversity can help to improve its resilience to shocks, as well as to promote its soil health and nutritional output (Kozicka et al. 2020). To realize diversity, farms should contains crops, livestock and/or forests to lower reliance on external output (Lemaire et al. 2014). Other integrated farm type is fish and crops farms, where fish effluents are used in irrigating the cultivated crops (van der Heijden et al. 2012). The resilience of farming system can be increased by the following:

6.1.1 Integrated Livestock and Crops Farms

In Egypt, many integrated farms exists, where livestock manure is as a fertilizer source applied to the cultivated crops in these farms. Eid et al. (2007) indicated that livestock and crops production is an excellent example of an integrated farm system, where feed crops and agricultural residues provide the feed for animals and, in turn, livestock manure is added to the soil to improve its fertility and reduce fertilizer use. Livestock manure is considered as a valuable input to soil for higher crop production and it improves soil health by inducing favorable physical, chemical and biological conditions (Alam et al. 2010). It is a cheap way for society to conserve the environment and reduce pollution (El Sheikha 2016).

It was reported that organic fertilizer can reduce salinity levels in the soil. Ding et al. (2020) indicated that application of organic manure, increased wheat yield grown in non-saline and saline soil under full irrigation by 18 and 11%, respectively. Whereas under deficit irrigation, wheat yield was increased by 11 and 17% under the addition organic manure, compared to without addition. They also added that soil N, P and K was increased in both soil types under both full and deficit irrigation treatments. Moreover, Zohry and Mohamadin (2020) indicated that application of cow manure and biofertilizer to soybean and sunflower increased its yield by 10 and 20%, respectively compared to application of cow manure only. Application of farmyard manure in addition to urea to potato improved its productivity by 79%, compared to application of urea only (Ahmed et al. 2015). Ali et al. (2011) stated that the application of farm manure increased canola yield, compared to no nitrogen application.

Compost is another type of farm wastes that could be used as organic fertilizer and an alternative soil amendment, which characterized by its slower nutrient release potential, compare to chemical fertilizer (Akanbi et al. 2007). Furthermore, adding compost improves soil fertility by increasing both the quantity and the quality of soil organic matter (Rivero et al. 2004). According to Eghball (2002), compost application in excess to crop requirements can last for several years in the soil since only a fraction of nitrogen and other nutrients in compost become available in the first year after application. Application of compost to the growing crops were studied in Egypt by El Sabagh et al. (2015), who stated that the application of compost increased canola yield, compared to no nitrogen application. Furthermore, Zohry and Mohamedin (2020) indicated that application of compost to soybean, sunflower and canola increased its yield, compared to application of cow manure.

6.1.2 Integrated Fish and Crops Farms

Integrated fish and crops farms attained similar sustainability as integrated livestock and crops farms (van der Heijden et al. 2012). Irrigating crops with farm fish effluents improves soil quality due to organic matter that exists in fish water (Elnwishy et al. 2006). Isitekhale and Adamu (2016) reported that only 25% of N and 20% of P of fish feed is recovered in harvested fish and the rest is accumulated in farm effluent. Incorporating these organic wastes into the soil helps to build up soil organic matter layers needed for steady supply of nutrients to the growing plants (Zieman et al 2007), increased the yield of cultivated crops and improve soil chemical properties, such as available phosphorus, water soluble potassium, calcium and magnesium (Ojobor and Tobih 2015). Thus, irrigation with fish farms effluents can substitute for 100% of fertilizer applied to winter crops (Zohry et al. 2020; Hefny et al. 2020) and for 50% of the applied fertilizer to summer crops (Zohry et al. 2020; Selim and shams 2019).

Positive effects of using fish farm effluent in irrigating potato, soybean, and onion were observed on their yield (Haquea et al. 2016). AbdEl Magid et al. (2018) reported that using fish farm effluent in irrigating onion and sugar beet increased their yields by

16% for both crops without adding fertilizers. The yield of tomato irrigated with fish water effluent increased in the first three harvestings (Castro et al. 2006). Furthermore, irrigation with fish farms effluents increased the yield of faba bean, lupine, soybean and sunflower by 13, 15, 9 and 3%, respectively (Zohry et al. 2020).

6.2 Increasing Crops Diversification

Crop diversification strategies simultaneously increase net crop production and also improve soil health by imparting interspecific interactions among different crop species both above and below ground, respectively (Rakshit et al. 2018). Above-ground diversification increases canopy heat and light capture, whereas belowground diversification assists in better utilization of water and soil macro- and micro-nutrients such as phosphorus (P), iron (Fe), and zinc (Zn) by increasing soil microbial load and activity (King and Hofmockel 2017). Increasing crops diversification can be done by the following climate-smart practices:

6.2.1 Cultivation of Three Crops Per year

The traditional crop cultivation in Egypt is the cultivation of a winter crop followed by a summer crop. However, cultivating three crops per year (winter, fall then summer crops or winter, early summer then late summer crops) instead of two crops per year increase crop diversification.

This system has many beneficial on land fertility, if legume crop was cultivated in between the winter crop and the summer crop. The major benefits resulted from this practice is the increase in land productivity, improve and sustain soil fertility and increase farmers' income (Sheha et al. 2014). It is also expected that this practice will increase soil water holding capacity. Zohry et al. (2017) compared between the effect of conventional crop sequence, namely wheat then maize with cultivation of three crops per year, namely maize, short-season clover, then wheat on wheat yield. Their results indicated that wheat productivity was increased by 16 and 47% in the first and second season, respectively when short-season clover followed maize, compared to the conventional crop sequence.

6.2.2 Implementing Intercropping Systems

Another form of biodiversity is the implementation of intercropping system instead of mono cropping. Intercropping systems can add another dimension to crop diversification and increase resilience to climate. Intercropping is a planting system with cultivation of two crops (main and secondary crops) resulting in increased light interception, root contact with more soil, microbial activity and deterrent to pests and weeds and final yield (Hauggaard-Nielsen et al. 2009). Intercropping alters the microclimate of the soil by changing soil temperature and moisture: it changes the pattern of dispersal through wind, rain, or a vector that inclusively benefits the intercropped plants in one way or another. It increases nitrogen and carbon content in the rhizospheric soil, and therefore those resource pools can be further used by successive crops (Zang et al. 2015). Xiong et al. (2013) reported increased iron (Fe) content in the rhizosphere of a peanut–maize intercropping system. The Fe enrichment enhanced the carbon and nitrogen metabolism and photosynthetic efficiency of the peanut crop as well as the resistance of both peanut and maize against various environmental stresses.

The efficiency of intercropping depends directly on proper management of production factors such as spatial arrangement of crops and planting density to reduce the competition for resources and increase the efficiency of the system (Porto et al. 2011). These factors, when properly managed, can bring ecological and economic benefits as a result of increasing production when compared to monoculture or chemical, physical and biological soil enrichment (Batista et al. 2016). It also increase farmers' income as a result of risks reduction from crop failure (Anderson 2007). Intercropping reduces water runoff and soil loss (Lithourgidis 2011), increases nutrient availability (Li et al. 2014), and improves soil quality (Cong et al. 2015). Furthermore, spatial arrangement and different pattern of roots exploit soil nutrients in this system minimize plants competition (Ijoyah and Fanen 2012). Several benefits have been attained from implanting intercropping systems in Egypt.

Biological Control of Insects and Diseases

One on the obstruction of climate variability is the outbreak of insects and diseases. Implementing intercropping system can attain biological control of insects and diseases. The companion crop in the intercropping system is grown closer to the main crop may produce chemical that suppress or repel pests (Ode 2006) and may feed and attract natural enemies (Parolin et al. 2012). Owing to the increasing environmental hazards and enhanced resistance to insecticides, using intercropping systems can minimize predator's infestation. In tomato, yield damage occur as result infestation with large numbers of insects. Khafagy (2015) indicated that intercropping aromatic plants with tomato reduces the infestation percentage with Tuta absoluta especially when geranium intercropped with tomato. It was also found that intercropping basil with cotton reduced all pest infestation and led to a 50% reduction in abundance of the pink bollworm, Pectinophora gossypiella Saunders (Schader et al. 2005). Furthermore, Hamd-Alla et al. (2014) stated that reduction in biological predators that attack maize plants was observed when cowpea was intercropped with maize, which positively affected maize yield. Smith et al. (2007) indicated that fusarium wilt, near wilt and root rot diseases cause partial to complete loss of pea crop and is capable of causing such damage to other crops as well. Gehad (2013) indicated that some medicinal plants could act as a natural disease control for pea, namely C. carvi, which increased pea seed yield, compared to sole pea cultivation. Maize intercropped with tomato proved to reduced pests and diseases that usually exist in tomato monoculture (Pino et al. 1994). Hao (2013) indicated that maize intercropping with tomato has control effect on powdery mildew that occurs in tomato plants.

Mitigation of Abiotic Stress

Some intercropping systems could modify the microclimate through shading. Maize intercropped with summer tomato (Mohamed et al. 2013) or sunflower intercropped with tomato (Kestha and El-Baz 2004) can modify the microclimate for tomato, where it protected the tomato fruits from sun damage in July and August (Mohamed et al. 2013). Abdelmageed et al. (2003) stated that the reproductive stage in tomato are more sensitive to high temperatures than the vegetative processes. Abdel (2006) indicated that intercropping sunflower with tomato resulted in mitigating heat stress through evaporative cooling and shading means, which improved fruit set and consequently yield quality. Furthermore, intercropping faba bean (Ibrahim et al. 2010) or wheat (Abd El-Zaher et al. 2013) with tomato in the winter provides warmth to tomato plants and extends its growing season for two months.

Implementing Crop Rotation

Implementing crop rotation is an important mean of crop diversification and to maintain soil sustainability. It is the cheapest and most effective way to improve productivity of the cultivated crops and soil fertility. Three principles are exist for crop rotation, diversity of crops in time and space at the field levels to increase the crops' innate capacity to suppress weeds, pests and disease; alternation of crops with different rooting depths; and each complete crop rotation should maintain or increase soil organic matter (Boincean and Dent 2019). Furthermore, in a crop rotation, different economically important crops are in succession on a particular field (Bruns 2012). It allows rotating a cereal crop with a legume crop to prevent soil deterioration resulted from continuous cultivation of cereal crops (Bennett et al. 2012). Jagadamma et al. (2008) indicated that the inclusion of soybean in a rotation with corn has been shown to increase soil aggregates stability, compared to corn continuous cultivation. Similarly, Stone and Silveira (2001) indicated that crop rotations composed of soybean and wheat resulted in the highest micro-porosity values and the lowest macro-porosity values, whereas the crop rotation composed of upland rice associated to calopogonium and common bean showed the highest macro-porosity value and the lowest micro porosity value. Soil organic matter content in all soil depths was increased and soil bulk density was reduced when wheat and sweet clover were in a crop rotation (Liu et al. 2005).

Crop rotations have been implemented in Egypt to sustainably increase yield of the cultivated crops and improve soil fertility. For example, Abou-Keriasha et al. (2012) implemented three cotton rotations, three years each, namely the rotation is the prevailing rotation in the surrounding areas. The second rotation was a low-intensive rotation, contained mono crops, and intercropping systems, in addition to three-crop sequence, mainly legumes. The third crop rotation was a highly intensive rotation, contained intercropping system for cotton and other intercropping systems, in addition to three-crop sequence, mainly legumes.

Their results indicated that the yield of the cultivated crops was increased in the second and third rotations, compared to its counterpart values under the prevailing crop rotation. Furthermore, soil organic matter content was increased by 22% in the low-intensive crop rotation, compared to the prevailing crop rotation and the values of soil available N, P, and K were increased by 41, 39 and 31%, respectively, compared to the prevailing crop rotation and the farmer's net revenue obtained from this rotation was increased by 20%, compared to its value under the prevailing rotation. In the highly-intensive crop rotation, soil organic matter content was increased by 38%, compared to the prevailing crop rotation. In addition, available N, P, and K values was increase by 43, 45, and 35%, respectively and the farmer's net revenue of this rotation was increased by 32%, compared to its value under the prevailing rotation.

6.3 *Increasing Resilience of Crops/Varieties*

Increasing the resilience of agricultural crops towards various biotic and abiotic stresses is a promising method for maximizing crop production under adverse conditions of drought, salinity and heat (Shen et al. 2015). Drought, salinity and heat are the most common abiotic stresses that faces crops.

6.3.1 Drought Stress

Areas under drought are very prone to soil erosion, which in turns leads to reduced soil fertility, soil nutrients, and soil productivity; all are important factors driving crop growth and productivity (Delgado et al. 2013). In these areas, sorghum is used as a drought-tolerant crop in marginal areas of Africa, where it has a vital role in livelihoods of millions of people (Westengen et al. 2014). Khatab et al. (2017) evaluated water stress tolerance on some sorghum genotypes and confirmed that some of them exhibited highly resistance for drought tolerance. El-Samnoudi et al. (2019) indicated that application of poultry manure and mulching to sorghum grown under deficit irrigation alleviated the effect of water stress and produced yield similar to the unstressed plants.

Furthermore, using plant growth-promoting rhizobacteria is found to be a better approach in the case of dry land agriculture as it introduces beneficial physical and chemical changes in the soil–plant system (Dubey et al. 2017). Talaat et al. (2015) found that foliar application of 0.1 mg L^{-1} 24-epibrassinolide (EBL) and 25 mg L^{-1} spermine (Spm) to two maize hybrids (Giza 10 and Giza 129) grown under drought stress conditions, resulted in increasing the drought tolerance. Furthermore,

Abdullah et al. (2015) found that negative effects of late-season drought on growth and yield of wheat could be mitigated by application of anti-transpirants, where it increased plant water use, rate of transpiration, stomatal conductance, leaf turgor, and photosynthesis, which collectively reduces water deficit or drought stress condition.

Furthermore, Abd Allah et al. (2021) indicated that spraying with 200 ppm potassium silicate for faba been intercropping system with sugar beet positively affected N, P, K, and Si contents, as well N, P, and K in the shoots and seeds of faba bean and in the soil under deficit irrigation, compared to no application. Whereas, similar trend was observed for sugar beets shoot and roots, as well as for the soil P content and sucrose percentage in sugar beet tubers, compared to no application. As a result, water equivalent ratio was increased by 8% under application of deficit irrigation and spraying with 200 ppm potassium silicate, compared to no application.

6.3.2 Salinity Stress

An improved salinity tolerance of crop plants by application of bio-fertilizers was reported by several investigation in Egypt. Gomaa and Gaballah (2004) reported that application of yeast to different maize hybrids alleviated the adverse effects of high levels of salinity and the plants accumulated more polyamines than those received no bio-fertilizer. Similar positive effect was observed in faba bean when the seeds was inoculated with yeast under saline conditions, which increased the percentage of germination, compared to seeds without inoculation (Gaballah and Gomaa 2004). Furthermore, Baddour et al. (2017) reported that spraying maize plants with ascorbic and proline, in addition to application of chicken manure increased maize yield under saline conditions.

6.3.3 Heat Stress

Heat stress negatively affects various plant development processes and eventually resulting in great yield loss (Grant et al. 2011). Quinoa is known as a heat tolerant crop. Hinojosa et al. (2019) reported that quinoa could adapt to high day/night temperatures by showing high plasticity in terms of several growth parameters, such as net photosynthesis, chlorophyll fluorescence, leaf greenness index, seed size and seed set. Similarly, pearl millet proved to have the ability to tolerate heat stress (Sage and Zhu 2011). Furthermore, it was documented that sorghum resiliency is expressed in being heat tolerant (Chiluwal et al. 2019). In Egypt, sorghum is widely cultivated as heat-tolerant crop in south Egypt to partially reduce reliance on wheat in bread making. In addition, forage sorghum is a popular crop used for feed in this region (Said et al. 2016).

6.4 *Early Warning System for Extreme Weather Events*

The concept of the early warning system is used in different contexts of risk management and is used as a tool for planning before extreme weather events (Basher 2006). In Egypt, Young et al. (2021) presented a practical approach of incorporating critical rainfall thresholds, historical flood data (crowd sourced) and ensemble precipitation forecast for forecasting extreme rainfall and flooding in Alexandria, North Egypt, which could help in improving decision-making especially in data-scarce regions or cities in the genesis of developing early warning system. Furthermore, Baldi et al. (2020) studied the spatio-temporal evolution of selected episodes of heavy rainfall trend over Sinai Peninsula, Egypt to improve early warning system. The study was based on the analysis of meteorological data from ground stations in Sinai and rainfall estimates derived by satellite images with the aim to help the decision-makers to plan the construction of flood water harvesting structures and flash flood protection works.

7 Conclusions

Variability in weather and climate inherently leads to the occurrence of extreme weather or climate events. Extreme events are expected to cause instability of crops yields and forming high risk to global crop production. Low precipitation events have been attributed to falls in crop productivity. Egypt have experience extreme weather events for the past few years. Several measured have been taken to limit the vulnerability of the agricultural system to climate variability. Reducing exposure to climate variability was done by implementing resilient farming systems, where integrated farming system were preformed, namely integrated livestock and crops, and integrated fish and crops. Increasing crops diversity was also done by cultivation of three crops per year, implementing intercropping systems and crop rotations. Furthermore, cultivation of resilient crops tolerant to drought, salinity and heat stresses was done, namely quinoa, sorghum and millet are used to reduce the sensitivity of the agriculture systems. Using early warning system for extreme weather events was also a very useful procedure to overcome its dreadful effects of climate variability and to increase adaptive capacity.

References

Abd Allah AMM, Mohamed MS, Noreldin T (2021) Water deficiency relief by application of potassium silicate to faba bean intercropped with sugar beet in sandy soil. Zagazig J Agric Res 48(1):1–18

Abd El-Zaher ShR, Shams AS, Mergheny MM (2013) Effect of intercropping pattern and nitrogen fertilization on intercropping wheat with tomato. Egypt J Appl Sci 28(9):474–489

Abdel CG (2006) Improvement of tomato fruit-set under natural high temperature: I. Intercropping tomato with sunflower or corn. J Dohuk Univ 9(2):2–16
AbdEl Magid HAA, Hala HAA, Mohamed AMM (2018) Economic study the efficiency of water resource usage (Case Study). Alex J Agric Sci 63(3):149–155
Abdelmageed AH, Gruda N, Geyer B (2003) Effect of high temperature and heat shock on tomato (Lycopersiconesculentum Mill.) genotypes under controlled conditions. In: Conference on international agricultural research for development, Göttingen, October 8–10
Abdullah AS, Aziz MM, Siddique KHM, Flower KC (2015) Film antitranspirants increase yield in drought stressed wheat plants by maintaining high grain number. Agric Water Manag 159:11–18
Abou-Keriasha MA, Mohamed WMA, Eiasa NM, Kamel AS (2012) Intensive crop rotations to improve agricultural production in Middle Egypt. Egypt J Agric Res 90(4):427–443
Adamgbe EM, Ujoh F (2013) Effect of variability in rainfall characteristics on maize yield in Gboko. Nigeria J Environ Prot 4:881–887
Ahmed AA, Zaki MF, Shafeek MR, Helmy YI, Abd El-Baky MMH (2015) Integrated use of farmyard manure and inorganic nitrogen fertilizer on growth, yield and quality of potato (Solanum tuberosum L.). Int J Curr Microbiol App Sci 4(10):325–349
Akanbi WB, Adebayo TA, Togun OA, Adeyeye AS, Olaniran OA (2007) The use of compost extract as foliar spray nutrient source and botanical insecticide in Telfairia occidentalis. World J Agric Sci 3(5):642–652
Al Zayed IS, Ribbe L, Al Salhi A (2013) Water harvesting and flashflood mitigation-wadi watier case study (South Sinai, Egypt). Int J Water Resour Arid Environ 2:102–109
Alam M, Siddiqua A, Chowdhury M, Prodhan M (2010) Nodulation, yield and quality of soybean as influenced by integrated nutrient management. J Bangladesh Agric Univ 7:229–234
Ali ME, Fathi AI, Mohamed OH, El-Edfawy YM (2011) Response of canola productivity and quality to bio- organic and inorganic N fertilizers. J Soil Sci Agric Eng Mansoura Univ 2(12):1255–1272
Almeselmani M, Deshmukh PS, Sairam RK (2009) High temperature stress tolerance in wheat genotypes: role of antioxidant defense enzymes. Acta Agron Hungar 57:1–14
Altieri MA, Nicholls CI, Henao A, Lana MA (2015) Agroecology and the design of climate change-resilient farming. Agron Sustain Dev 35:869–890
Anderson RL (2007) Managing weeds with a dualistic approach of prevention and control: a review. Agron Sustain Dev 27:13–18
Baddour AG, Rashwan EM, El-Sharkawy TA (2017) Effect of organic manure, antioxidant and proline on corn (Zea mays L.) grown under saline conditions. Env Biodiv Soil Secur 1:203–217
Baldi M (2019) Climate change, water availability, and cultural heritage in Egypt. In: Baldi M, Vittozzi GC (eds) Sciences and technologies applied to cultural heritage—STACH 1. Cnr Edizioni, Roma, Italy. https://iiccairo.esteri.it/iic_ilcairo/it/istituto/centro-archeologico/link-utili
Baldi M, Amin D, Al Zayed IS, Dalu G (2020) Climatology and dynamical evolution of extreme rainfall events in the Sinai Peninsula—Egypt. Sustainability 12:6186. https://doi.org/10.3390/su12156186
Basher R (2006) Global early warning systems for natural hazards: systematic and people-centered. Philos Trans Ser A Math Phys Eng Sci 364(1845):2167–2182
Bathiany S, Dakos V, Scheffer M, Lenton TM (2018) Climate models predict increasing temperature variability in poor countries. Sci Adv 4 eaar5809
Batista TMV, Bezerra FN, Silva IN, Da Silva ML, De Oliveira EQ, Junior APB (2016) Agronomic efficiency of the intercropping of arugula with carrot under different population combinations. Rev Caatinga Mossoró 29(1):76–84
Ben-Asher J, Garcia AGY, Hoogenboom G (2008) Effect of high temperature on photosynthesis and transpiration of sweet corn (Zeamays L. var. rugosa). Photosynthetica 46:595–603
Bennett AJ, Bending GD, Chandler D, Hilton S, Mills P (2012) Meeting the demand for crop production: the challenge of yield decline in crops grown in short rotations. Biol Rev 87:52–71
Boincean B, Dent D (2019) Crop rotation. In: Farming the black earth. Springer Nature Switzerland AG. https://doi.org/10.1007/978-3-030-22533-9_5

Bruns HA (2012) Concepts in crop rotations. In: Aflakpui G (ed) Agricultural science. In Tech. ISBN: 978-953-51-0567-1. http://www.intechopen.com/books/agricultural-science/conceptsin-crop-rotation

Bullock JM, Dhanjal-Adams KL, Milne A, Oliver TH, Todman LC, Whitmore AB, Pywell RF (2017) Resilience and food security: rethinking an ecological concept. J Ecol 105:880–884

Castro RS, Azevedo CMB, Bezerra-Net F (2006) Increasing yield of cherry tomato using fish effluent as irrigation water in Northeast Brazil. Sci Hortic 110(1):44–506

Chiluwal A, Bheemanahalli R, Kanaganahalli V, Boyle D, Perumal R, Pokharel M, Oumarou H, Jagadish SVK (2019) Deterioration of ovary plays a key role in heat stress-induced spikelet sterility in sorghum. Plant Cell Environ 43:448–462

Chaves MM, Zarrouk O, Francisco R, Costa JM, Santos T, Regalado AP, Rodrigues ML, Lopes CM, Costa J, Lopes C (2010) Grapevine under deficit irrigation: hints from physiological and molecular data. Ann Bot 105:661–676

Cong WF, Hoffland E, Li L, Six J, Sun JH, Bao XG (2015) Intercropping enhances soil carbon and nitrogen. Glob Chang Biol 21:1715–1726. https://doi.org/10.1111/gcb.12738

Davis KF, Chhatre A, Rao ND, Singh D, DeFries R (2019) Sensitivity of grain yields to historical climate variability in India. Environ Res Lett 14(614):064013

Delgado JA, Nearing MA, Rice CW (2013) Conservation practices for climate change adaptation. Adv Agron 121:47–115

Delpla I, Jung AV, Baures E, Clement M, Thomas O (2009) Impacts of climate change on surface water quality in relation to drinking water production. Environ Int 35:1225–1233

Ding Z, Kheir AMS, Ali MGM, Ali OAM, Aly, Abdelaal IN, Lin X, Zhou Z, Wang B, Liu B, He Z (2020) The integrated effect of salinity, organic amendments, phosphorus fertilizers, and deficit irrigation on soil properties, phosphorus fractionation and wheat productivity. 10:2736. https://doi.org/10.1038/s41598-020-59650-8

Dubey RK, Tripathi V, Edrisi SA, Bakshi M, Dubey PK, Singh A, Verma JP, Singh A, Sarma BK, Rakshit A et al (2017) Role of plant growth promoting microorganisms in sustainable agriculture and environmental remediation. In: Singh HB, Sharma B, Kesawani C (eds) Advances in PGPR research. CABI Press, Boston, MA, USA

Eghball B (2002) Soil properties as influenced by phosphorus- and nitrogen-based manure and compost applications. Agron J 94:128–135

Eid HM, El-Marsafawy SM, Ouda SA (2007) Assessing the economic impacts of climate change on agriculture in Egypt: a Ricardian approach. Policy research working paper no. WPS4293. The World Bank Development Research Group

El Sabagh A, Abd Omar E, Saneoka H, Barutçular C (2015) Evaluation of agronomic traits of canola (Brassica napus L.) under organic, bio- and chemical fertilizers. DUFED 4(2):59–67

El Sheikha AF (2016) Mixing manure with chemical fertilizers, why? and what is after? Nutr Food Technol 2(1):112. https://doi.org/10.16966/2470-6086

Elahia E, Khalid Z, Taunic MZ, Zhanga H, Lironga X (2021) Extreme weather events risk to crop-production and the adaptation of innovative management strategies to mitigate the risk: a retrospective survey of rural Punjab, Pakistan. Technovation 102255

Elnwishy N, Salh M, Zalat S (2006) Combating desertification through fish farming: the future of drylands. In: Proceedings of the international scientific conference on desertification and drylands research, Tunisia, 19–21 June, UNESCO

El-Samnoudi IM, Ibrahim AM, Abd El Tawwab AR, Abd E-MA (2019) Combined effect of poultry manure and soil mulching on soil properties, physiological responses, yields and water-use efficiencies of sorghum plants under water stress. Commun Soil Sci Plant Anal 50(20):2626–2639. https://doi.org/10.1080/00103624.2019.167144

Gaballah MS, Gomaa AM (2004) Performance of faba bean varieties grown under salinity stress and biofertilized with yeast. J Appl Sci 4:93–99

Gehad MMJ (2013) Effect of intercropping of pea with some medicinal plants on microbial community of soil, damping-off and downy mildew diseases, under Beheira governorate conditions. Plant Prot Path Mansoura Univ 4(7):625–641

Gitz V, Meybeck A (2012) Risks, vulnerabilities and resilience in a context of climate change. In: Building resilience for adaptation to climate change in the agriculture sector. FAO, Rome, Italy, no 23, p 19

Gomaa AM, Gaballah MS (2004) Changes in compatible solutes of some maize varieties grown in sandy soil and biofertilized with Rhodotorula glutinis under saline conditions [online]. Zu finden in http://www.bodenkunde2.uni-freiburg.de/eurosoil/abstracts/id105_Gaballah_full.pdf

Gornall J, Betts R, Burke E, Clark R, Camp J et al (2010) Implications of climate change for agricultural productivity in the early twenty-first century. Philos Trans R Soc Lond B Biol Sci 365(554):2973–2989

Grant RF, Kimball BA, Conley MM, White JW, Wall GW, Ottman MJ (2011) Controlled warming effects on wheat growth and yield: field measurements and modeling. Agron J 103(6):1742–1754. https://doi.org/10.2134/agronj2011.0158

Hamd-Alla WA, Shalaby EM, Dawood RA, Zohry AA (2014) Effect of cowpea (Vigan sinensis L.) with maize (Zea mays L.) intercropping on yield and its components. Inter Scho Sci Res Innov 8(11):1170–1176

Hao W (2013) Control effect of tomato and maize intercropping against tomato powdery mildew. Plant Dis Pests 4(2):22–24

Haquea MM, Belton B, Mehedi AMD, Golzar AA, Rushna AMd (2016) Reuse of fish farm sediment as fertilizer for fodder grass production in Bangladesh: potential for sustainable intensification and improved nutrition. Agric Ecosyst Environ 216:226–236

Hauggaard-Nielsen H, Mundus S, Jensen ES (2009) Nitrogen dynamics following grain legumes and subsequent catch crops and the effects on succeeding cereal crops. Nutr Cycl Agroecosyst 84(3):281–291. https://doi.org/10.1007/s10705-008-9242-7

Hefny Y, Kassem E, El-Shenawy I (2020) Effect of irrigation with fish farm effluent on two legume crops interplanted with orange trees in a sandy soil. Mor J Agri Sci 1(6):291–300

Hegerl GC, Zwiers FW, Braconnot P, Gillett NP, Luo Y, Orsini JAM, Nicholls N, Hoogenboom G (2000) Contribution of agrometeorology to the simulation of crop production and its applications. Agric for Meteorol 103:137–157

Hinojosa L, Matanguihan JB, Murphy KM (2019) Effect of high temperature on pollen morphology, plant growth and seed yield in quinoa (Chenopodium quinoa Willd.). J Agro Crop Sci 205:33–45. https://doi.org/10.1111/jac.12302

Hoffmann U (2013) Section B: Agriculture: a key driver and a major victim of global warming. In: Lead article, in: Chapter 1, in Hoffmann, pp 3–5

Ibrahim S, Shaaban M, Gendy EK (2010) Intercropping faba bean with tomato. Egypt J Appl Sci 25(6A):167–181

Ijoyah MO, Fanen FT (2012) Effects of different cropping pattern on performance of maize-soybean mixture in Makurdi, Nigeria. Sci J Crop Sci 1(2):39–47

IPCC (Inter-Governmental Panel on Climate Change) (2007) Contribution of working group II to the fourth assessment report of the intergovernmental panel on climate variability. Cambridge University Press, Cambridge, p 60

IPCC (Inter-Governmental Panel on Climate Change) (2011) Climate change: impacts, adaptation, and vulnerability. In: McCarthy JJ, Canziani OF, Leary NA, Dokken DJ, White KS (eds) Contribution of working group II to the third assessment report of the intergovernmental panel on climate change. Cambridge University Press, Cambridge

IPCC (2014) Climate change 2014: impacts, adaptation, and vulnerability. Part A: Global and Sectoral aspects. Summary for policymakers. Contribution of working group II to the fifth assessment report of the intergovernmental panel on climate change (IPCC). Cambridge University Press, Cambridge and New York, NY, pp 1–32

Isitekhale HH E, Adamu B (2016) Effects of effluents on soil chemical Properties in forest Derived savanna Transition. IOSR J Environ Sci Toxicol Food Technol 10(6)11:30–34

Jagadamma S, Lal R, Hoeft RG, Nafziger ED, Adee EA (2008) Nitrogen fertilization and cropping system impacts on soil properties and their relationship to crop yield in the central Corn Belt, USA. Soil Tillage Res 98:120–129. https://doi.org/10.1016/j.still.2007.10.008

Jagadish SVK, Craufurd PQ, Wheeler TR (2007) High temperature stress and spikelet fertility in rice (Oryza sativa L.). J Exp Bot 58(7):1627–1635

Kestha MM, El-Baz MG (2004) Studies on sunflower-tomato intercropping. In: Proceedings of the 16th international sunflower conference, Fargo, ND USA, pp 514–518

Khafagy IF (2015) The role of some aromatic plants intercropping on tuta absoluta infestation and the associated predators on tomato. Egypt J Plant Protec Res 3(2):37–53

Khalil AA, Hassanein MK (2016) Extreme weather events and negative impacts on Egyptian agriculture. Int J Adv Res 4(12):1843–1851

Khatab IA, El-Mouhamady AA, Abdel-Rahman HM, Farid MA, El-Demardash IS (2017) Agro-morphological and molecular characterization of sorghum (Sorghum vulgare L.) for water stress tolerance. Int J Curr Res Biosci Plant Biol 4:37–55

King AE, Hofmockel KS (2017) Diversified cropping systems support greater microbial cycling and retention of carbon and nitrogen. Agric Ecosyst Environ 240:66–76

Kosova K, Vitamvas P, Prasil IT, Renaut J (2011) Plant proteome changes under abiotic stress-contribution of proteomics studies to understanding plant stress response. J Proteome 74:1301–1322. https://doi.org/10.1016/j.jprot.2011.02.006

Kozicka M, Gotor E, Ocimati W, Jager T, Kikulwe E, Groot JCJ (2020) Responding to future regime shifts with agrobiodiversity: a multi-level perspective on small-scale farming in Uganda. Agric Syst (183):102864

Kumar NM, Murthy CS, SeshaSai MVR, Roy PS (2009) On the use of standardized precipitation index (SPI) for drought intensity assessment. Meteorol Appl. https://doi.org/10.1002/met.136

Lemaire G, Franzluebbers A, de Carvalho PCF, Dedieu B (2014) Integrated crop-livestock systems: strategies to achieve synergy between agricultural production and environmental quality. Agric Ecosyst Environ 190:4–8. https://doi.org/10.1016/j.agee.2013.08.009

Li Y, Zhang W, Ma L, Wu L, Shen J, Davies WJ, Oenema O, Zhang F, Dou Z (2014) An analysis of China's grain production: looking back and looking forward. Food Energy Sec 3:19–32

Lithourgidis AS (2011) Annual intercrops: an alternative pathway for sustainable agriculture. Aust J Crop Sci 5(4):396–410

Liu X, Herbert SJ, Hashemi AM, Zhang X, Ding G (2005) Effects of agricultural management on soil organic matter and carbon transformation—a review. Plant Soil Environ 56:531–543

Lobell DB, Monasterio O (2007) Impact of day versus night temperature on spring wheat yield: a comparison of empirical and CERES model predictions in three locations. Agron J 99:469–477

Lobell DB, Hammer GL, McLean G, Messina C, Roberts MJ, Schlenker W (2013) The critical role of extreme heat for maize production in the United States. Nat Clim Change 3:497–501

Mahmood S, Wahid A, Javed F, Basra SMA (2010) Heat stress effects on forage quality characteristics of maize (Zea mays) cultivars. Int J Agric Biol 12:701–706

Manneh B, Kiepe P, Sie M, Ndjiondjop M, Drame NK, Troare K, Rodenburg J, Somado Mbilinyi BP, Tumbo SD, Rwehumbiza KD (2010) Economics of climate change for agriculture sector in Tanzania. http://www.suanet.ac.tz/ccaa

Matchett JR, Berlow EL, Ostoja SM, Knapp RA, McKenny HC, Brooks ML, Stark PB (2015) Detecting the effects of rare stressors on rare species: is Yosemite toad breeding influence by packs tock use? Nat Sci Rep 5:10702. https://doi.org/10.1038/srep10702

Meuwissen MPM, Feindt PH, Spiegel A, Termeer CJ, Mathijs E, de Mey Y, Finger R, Balmann A, Wauters E, Urquhart J et al (2019) A framework to assess the resilience of farming systems. Agric Syst 176

Moawad MB, Abdelaziz AO, Mamtimin B (2016) Flash floods in the Sahara: a case study for the 28 January 2013 flood in Qena, Egypt. Geomat Nat Hazards Risk 7(1):215–236. https://doi.org/10.1080/19475705.2014.885467

Mohamed W, Ahmed NR, Abd El-Hakim WM (2013) Effect of intercropping dates of sowing and N fertilizers on growth and yield of maize and tomato. Egypt J App Sci 28(12B):625–644

Motha RP (2011) The impact of extreme weather events on agriculture in the United States. In: Challenges and opportunities in agrometeorology. https://doi.org/10.1007/978-3-642-19360-6_30

Nelson D, Adger W, Brown K (2007) Adaptation to environmental change: contributions of a resilience framework. Annu Rev Environ Resour 32:395–419
Ode PJ (2006) Plant chemistry and natural enemy fitness: effects on herbivore and natural enemy interactions. Annu Rev Entomol 51:163–185
Ojobor SA, Tobih FO (2015) Effects of fish farm effluent and inorganic fertilizer on Amaranthus yield and soil chemical properties in Asaba, Delts State Nigeria. J Agric Environ Sci 4(1):237–244
Oshino T, Miura S, Kikuchi S, Hamada K, Yano K, Watanabe M, Higashitani A (2011) Auxin depletion in barley plants under high temperature conditions represses DNA proliferation in organelles and nuclei via transcriptional alterations. Plant Cell Environ 34:284–290. https://doi.org/10.1111/j.1365-3040.2010.02242.x
Parolin P, Bresch C, Poncet C, Desneux N (2012) Functional characteristics of secondary plants for increased pest management. Int J Pest Manag 58(4):369–377. https://doi.org/10.1080/09670874.2012.734869
Pino M, De-Los A, Bertoh M, Espinosa R (1994) Maize as a protective crop for tomato in conditions of environmental stress. Cult Trop 15:60–63
Porter JR, Xie L, Challinor AJ, Cochrane K, Howden M, Iqbal MM, Lobell DB, Travasso MI (2014) Food security and food production systems. Climate change 2014: impacts, adaptation and vulnerability. Working group II contribution to the IPCC 5th assessment report, Geneva, Switzerland
Porto VCN, Bezerra NF, Lima JSS, Barros JAP, Moreira JN (2011) Combination of lettuce and rocket cultivars in two cultures intercropped with carrots. Horti Bras 29:404–411
Powell JP, Reinhard S (2016) Measuring the effects of extreme weather events on yields. Weather Clim Extrem 12:69–79
Prasad PVV, Djanaguiraman M (2014) Response of floret fertility and individual grain weight of wheat to high temperature stress: sensitive stages and thresholds for temperature and duration. Funct Plant Biol 41:1261–1269. https://doi.org/10.1071/FP14061
Rakshit A, Sarkar B, Abhilash PC (2018) Soil amendments for sustainability: challenges and perspectives. CRC Press, Boca Raton, FL
Rivero C, Tchirenje L, Ma Q, Martinez G (2004) Influence of compost on soil organic matter quality under tropical conditions. Geoderma 123:355–361
Rosenzweig C, Iglesias A, Yang XB, Epstein PR, Chivian E (2001) Climate change and extreme weather events—implications for food production, plant diseases, and pests. Glob Change Hum Health 2(2):90–104
Sage RF, Zhu XG (2011) Exploiting the engine of C4 photosynthesis. J Exp Bot 62:2989–3000
Said AS, Zohry AA, Ouda S (2016) Unconventional solution to increase crop production under water scarcity. In: Major crops and water scarcity in Egypt. Springer Publishing House, pp 99–114
Saleh SM, Heggi MAM, Abdrabbo MA, Farag AA (2017) Heat waves investigation during last decades in some climatic regions in Egypt. Egypt J Agric Res 95(2):863–889
Sami Ul Din M, Ahmad I, Hussain N, Ahmad A, Wajid A, Khaliq T, Mubeen M, Imran M, Ali A, Akram R, Amanet K, Saleem M, Nasim W (2019) Agronomic cropping systems in relation to climatic variability. In: Hasanuzzaman M (ed) Agronomic crops. Springer Nature Singapore Pte Ltd. https://doi.org/10.1007/978-981-32-9151-5_5
Satake T, Yoshida S (1978) High temperature-induced sterility in indica rices at flowering. Jpn J Crop Sci 47(1):6–17
Schader C, Zaller JC, Köpke U (2005) Cotton-Basil intercropping: effects on pests, yields and economical parameters in an organic field in Fayoum, Egypt. Biol Agric Horticult 23:1. https://doi.org/10.1080/01448765.2005.9755308
Schlenker W, Roberts JM (2009) Nonlinear temperature effects indicate severe damages to U.S. crop yields under climate change. PNAS 106(37):15594–15598. https://doi.org/10.1073/pnas.0906865106
Selim MAF, Shams AS (2019) Maximizing efficiency of land and water utilization and profitability of interplanting maize with mandarin trees using irrigation with fish waste water between sandy soil and drip irrigation conditions. Middle East J Agric Res 8(04):1240–1252

Shah N, Paulsen G (2003) Interaction of drought and high temperature on photosynthesis and grain-filling of wheat. Plant Soil 257:219–226

Sheha AM, Ahmed NR, Abou-Elela AM (2014) Effect of crop sequence and nitrogen levels on rice productivity. Ann Agric Sci 52(4):451–460

Shen H, Zhong X, Zhao F, Wang Y, Yan B, Li Q, Chen G, Mao B, Wang J, Li Y, Xiao G, He Y, Xiao H, Li J, He Z (2015) Overexpression of receptor-like kinase ERECTA improves thermotolerance in rice and tomato. Nat Biotechnol 33(9):996–1003

Shiferaw B, Prasanna BM, Hellin J, Bänziger M (2011) Crops that feed the world: past successes and future challenges to the role played by maize in global food security. Food Secur 3:307–311

Sivakumar MVK, Das HP, Brunini O (2005) Impacts of present and future climate variability and change on agriculture and forestry in the arid and semi-arid tropics. Clim Change 70(1–2):31–72

Sivakumar MVK, Motha RP, Wilhite DA, Qu JJ (2011) Towards a compendium on national drought policy: proceedings of an expert meeting, July 14–15, Washington DC, USA. AGM-12/WAOB-2011. World Meteorological Organization, Geneva, 135 pp

Solomon S, Qin D, Manning M, Marquis M, Averyt K, Tignor MMB, Miller LH, Chen Z (eds) (2007) Climate change: the physical science basis. Contribution of working group I to the fourth assessment report of the intergovermental panel on climate change. Cambridge University Press, New York, NY

Smith SM (2012) Food crisis in the horn of Africa: progress report, July 2011–2012 (Oxfam International)

Smith P, Martino D, Cai Z, Gwary D, Janzen H, Kumar P, McCarl B, Ogle S, O'Mara F, Rice C, Scholes B, Sirotenko O, Howden M, McAllister T, Pan G, Romanenkov V, Schneider U, Towprayoon S (2007) Policy and technological constraints to implementation of greenhouse gas mitigation options in agriculture. Agr Ecosyst Environ 118:6–28

Stone LF, Silveira PM (2001) Effects of soil tillage systems and crop rotations on soil porosity and bulk density. Rev Bras Ciênc Solo 25(2):345–351

Sugiyama M, Akashi O, Wada K, Kanudia A, Li J, Weyant J (2014) Energy efficiency potentials for global climate change mitigation. Clim Chang 123(3–4):397–411

Sweet SKD, Wolfe WA, De Gaetano R, Bennerc (2017) Anatomy of the 2016 drought in the Northeastern United States: implications for agriculture and water resources in humid climates. Agric Forest Meteorol 247:571–581

Taha A (2012) Effect of climate change on maize and wheat grown under fertigation treatments in newly reclaimed soil. PhD thesis, Tanta University, Egypt

Talaat NB, Shawky BT, Ibrahim AS (2015) Alleviation of drought-induced oxidative stress in maize (Zea mays L.) plants by dual application of 24-epibrassinolide and spermine. Environ Exp Bot 113:47–58

Tian L, Yuan S, Quiring SM (2018) Evaluation of six indices for monitoring agricultural drought in the south-central United States. Agric Meteorol 249:107–119

van der Heijden PGM (2012) Water use at integrated aquaculture agriculture farms. Experiences with limited water resources in Egypt. Wageningen University and Research Centre for Development Innovation

van der Velde M, Tubiello FN, Vrieling A et al (2012) Impacts of extreme weather on wheat and maize in France: evaluating regional crop simulations against observed data. Clim Change 113:751–765. https://doi.org/10.1007/s10584-011-0368-2

van der Wiel K, Bintanja R (2021) Contribution of climatic changes in mean and variability to monthly temperature and precipitation extremes. Commun Earth Environ. https://doi.org/10.1038/s43247-020-00077-4

Westengen OT, Okongo MA, Onek L, Berg T, Upadhyaya H, Birkeland S, Khalsa SDK, Ringa KH, Stenseth NC, Brysting AK (2014) Ethnolinguistic structuring of sorghum genetic diversity in Africa and the role of local seed systems. Proc Natl Acad Sci USA 111:14100–14105

Wilhite DA, Svoboda MD, Hayes MJ (2005) Monitoring drought in the United States: status and trends. In: Boker VK, Cracknell AP, Heathcoe RL (eds) Monitoring and predicting agricultural drought—a global study. Oxford University Press, New York, pp 121–131

Xiong H, Shen H, Zhang L, Zhang Y, Guo X, Wang P, Duan P, Ji C, Zhong L, Zhang F, Zuo Y (2013) Comparative proteomic analysis for assessment of the ecological significance of maize and peanut intercropping. J Proteomics 78:447–460

Yadav SS, Redden R, McNeil DL, Patil SA (2010) Climate change and management of cool season grain legume crops. Clim Change Manag Cool Seas Grain Legum Crop 2010:1–460

Yehia AG, Fahmy KM, Mehany MAS, Mohamed GG (2017) Impact of extreme climate events on water supply sustainability in Egypt: case studies in Alexandria region and Upper Egypt. J Water Clim Change 8(3):484–494

Young A, Bhattacharya B, Zevenbergen C (2021) A rainfall threshold-based approach to early warnings in urban data-scarce regions: A case study of pluvial flooding in Alexandria, Egypt. J Flood Risk Manag https://doi.org/10.1111/jfr3.12702

Zang H, Yang X, Feng X, Qian X, Hu Y, Ren C, Zeng Z (2015) Rhizodeposition of nitrogen and carbon by mungbean (Vigna radiata L.) and its contribution to intercropped Oats (Avena nuda L.) Plos One 10(3):e0121132. https://doi.org/10.1371/journal.pone.0121132

Zieman DA, Walsh WA, Saphore, Berneth KF (2007) A survey of water qualities characteristics of effluent from Hawaian Facilities. J World Aquac Soc 23(3):180–191

Zhao F, Zhang D, Zhao Y, Wang W, Yang H, Tai F (2016) The difference of physiological and proteomic changes in maize leaves adaptation to drought, heat, and combined both stresses. Front Plant Sci 7:1471. https://doi.org/10.3389/fpls.2016.01471

Zohry A, Mohamadin AA (2020) Chemical versus nonchemical fertilizer sources to enhance the yield of three edible oil crops in East El-Owinate region in Egypt. J Soils Crops 30(2):191–199

Zohry AA, Ouda S, Hamd-Alla W, Shalaby E (2017) Evaluation of different crop sequences for wheat and maize in sandy soil. Acta Agric Slov 109(2):383–392

Zohry A, Hefny Y, Ouda S (2020) Evaluation of different crop sequences and water qualities treatments on orange yield under intercropping conditions in sandy soil. In: Proceedings of the 16th international conference on crop science, 16th of October, Cairo, Egypt. 16th of October, Cairo, Egypt, pp 315–340

Chapter 6
Climate-Resilient Crops

Abd El-Hafeez Zohry and Samiha Ouda

Abstract Extreme weather events such as droughts, heat waves and floods have serious implications on food security, especially in rural communities with high populations of small-scale farmers. Climate-resilient crops and/or crop varieties have enhanced tolerance to biotic and abiotic stresses. They are intended to maintain or increase crop yields under stressful conditions and thereby provide a mean of adapting to diminishing crop yields in the face of droughts, higher average temperatures and other climatic conditions. It is adapted to adverse soil and climatic conditions in places where most agriculture is marginal. Examples of climate-resilient crops are quinoa and pearl millet. Both crops were documented for their ability to tolerant heat, salinity and water stresses. Quinoa was introduced in Egypt in 2011, and it can be used in bread making. On the other hand, pearl millet have been cultivated in Egypt as a forage crop. Furthermore, sorghum is also considered as a climate-resilient crop. It is more common in south Egypt, where it cultivated for its grain, which is used in bread making. In this chapter, we reviewed the resiliency of quinoa, pearl millet and sorghum crops to climate variability and change and its contributions in increasing food security worldwide.

Keywords Quinoa · Pearl millet · Sorghum · Heat stress · Salinity stress · Water stress

1 Introduction

Agriculture and food production are highly vulnerable to climate variability. Extreme weather events such as droughts, heat waves and flooding have serious implications on food security, especially in rural communities with high populations of small-scale

A. E.-H. Zohry (✉)
Crops Intesefication Research Department, Field Crops Research Institute; Agricultural Research Center, Giza, Egypt

S. Ouda
Water Requirements and Field Irrigation Research Department, Soils, Water and Environment Research Institute; Agricultural Research Center, Giza, Egypt

S. Ouda and A. E.-H. Zohry, *Climate-Smart Agriculture*,
https://doi.org/10.1007/978-3-030-93111-7_6

farmers. To cope with climate variability, farmers need to modify farm management practices, such as adjusting planting time, supplementing irrigation (when possible), implementing intercropping systems, adopting conservation agriculture, and planting more climate-resilient crops or varieties. The challenge to feed the ever-growing population with a healthy balanced diet and the threats faced by agricultural crops due to changing climate highlight the immediate requirement to exploit the beneficial attributes of the climate-resilient crops (Amadou et al. 2013). Cultivation of climate-resilient crops and/or varieties is considered one of the most proactive measures that can be taken to cope with food insecurity caused by unpredictable weather patterns (Smith 2012).

Climate-resilient crops were defined by Dhankher and Foyer (2018) as "both crops and crop varieties with enhanced tolerance to biotic and abiotic stresses, which have the ability to maintain or increase crop yields under stress conditions such as drought, flooding, heat, chilling, freezing and salinity". Climate-resilient crops and crop varieties have enhanced tolerance to biotic and abiotic stresses (Smith 2012). They are intended to maintain or increase crop yields under stress conditions and thereby provide a means of adapting to diminishing crop yields in the face of droughts, higher average temperatures and other climatic conditions (Gollin et al. 2005). Adoption of climate-resilient crops, such as early-maturing crop varieties, heat-tolerant varieties, drought-tolerant crops, and crops or varieties with enhanced salinity tolerance can help farmers to better cope with climate shocks (Lin 2011).

In the past few decades, few crop species were available to produce food worldwide. For example, wheat (Triticum aestivum), maize (Zea mays L.), rice (Oryza sativa L.) and soybeans (Glycine max) are the major crops in food production worldwide. However, the production of these major crops has been threatened by the increasing incidence of climate variability in many countries (Massawe et al. 2016). On the contrary, an abundance of genetic resources can be utilized to meet the growing food demand, and it possess resilience under harsh weather conditions (Dida and Devos 2006). Exploring the benefits and the opportunities of growing these resilient crops would contribute in diversifying agricultural systems and thereby improve food security.

Examples of climate-resilient crops are quinoa, pearl millet and sorghum. The quinoa plants are reported to be tolerant to heat stress (Hinojosa et al. 2019), drought (Garcia et al. 2007) and can tolerate water stress (Wu et al. 2016). Quinoa was introduced in Egypt in 2011, where the first field experiment was implemented. Similarly, pearl millet can tolerate heat stress (Sage and Zhu 2011), salinity stress (Yakubu et al. 2010), and water stress (Sher et al. 2019). Pearl millet have been cultivated in Egypt as a forage crop. Furthermore, sorghum is also considered as a climate-resilient crop. It was documented that sorghum resiliency is expressed in being heat tolerant (Chiluwal et al. 2019), salinity tolerant (Shakeri and Emam 2017) and water stress tolerant (Surwenshi et al. 2010). It is more common in south Egypt to be cultivated for its grain and it is used in bread making.

In this chapter we reviewed the resiliency of quinoa, pearl millet and sorghum crops to climate variability and change that can contribute to increasing food security worldwide.

2 Advantage of Using Climate-Resilient Crops

Cultivation of drought and salt tolerant crops has the potential to enhance farm-level productivity and livelihoods in drought and salt prone areas (Kaya and Yazar 2016). There are three well-known recognized crops for its resiliency to extreme weather conditions. These crops are quinoa, millet and sorghum.

2.1 The Advantage of Using Quinoa as a Food Crop

Quinoa is considered as a halophytic grain crop (Ruiz et al. 2016) grown in various climatic zones (Jacobsen et al. 2012) and can be cultivated in low fertile soil (Vilche et al. 2003). At a global level, there are more than 6000 varieties of quinoa that are cultivated by farmers (Rojas et al. 2015). Those varieties can be classified into five main categories or ecotypes, according to their adaptation to specific agro-ecological conditions in major production areas (Bazile et al. 2015). It is adapted to adverse soil and climatic conditions in places, where most agriculture is marginal (Garcia et al. 2015).

3 Nutritional Advantage of Quinoa

In 2013, the United Nations declared the International Year of Quinoa. FAO organization defined quinoa to be the only plant food that provides all essential amino acids (Romero and Shahriari 2011). It has been selected by Food and Agriculture Organization of The United Nations (FAO 2013), as one of the crops destined to contribute to food security in this century. It has nutritional advantage over major cereals of the world. Quinoa seeds contained 16.5 g protein, 6.3 g fat, 3.8 g ash, 3.8 g crude fiber, 69.0 g carbohydrate, and energy amount of 399 kcal. All these nutrients are higher than what found in wheat, maize, rice and sorghum (Hulse et al. 1980; United States National Research Council).

Quinoa is classified within the super foods for its high content in protein. It also has lysine, which is considered the first limiting essential amino acid in cereals (Arendt and Zannini 2013). Quinoa seed has an extraordinary nutritional value for human, as it have an excellent balance of carbohydrates, lipids and protein for nutrition (Maradini-Filho et al. 2017). Quinoa protein has also contains lysine and sulfur amino acids (Escuredo et al. 2014), which meet and surpass the daily intake requirements that is recommended by the World Health Organization (Hirose et al. 2010). When it properly handled to remove the bitter saponins in the pericarp, quinoa grains has a mild flavor and can be consumed the same ways as cereal grains (Fuentes-Bazan et al. 2012).

Although quinoa grains do not contain gluten, it can be mixed with wheat flour in the preparation of bread with high nutritional value (Morita et al. 2001). Abdellatif (2018) indicated that the chemical composition of quinoa seeds can be successfully used as a food ingredient to develop new products with high nutritional and organoleptic properties. Substituting wheat flour with 20% quinoa resulted in elevating protein, fat and fiber percentages than that of wheat flour (Soliman et al. 2019). Furthermore, quinoa is cheap to produce due to its ability to adapt to different climate and soil conditions and its broad genetic variability (Molina-Montenegro 2013).

Processing of quinoa produces pearled quinoa, granules, flakes, flour, nodules, oil, pasta and starch (Francis et al. 2002). Fresh quinoa leaves and harvest chaff are quite attractive for sheep, goats and fish (Arendt and Zannini 2013) and quinoa leaves may be used for silage (Bazile et al. 2015). Residue from milling has a high nutritional content and can be used as animal feed. Low quality broken grains are used for poultry feed whereas stalks, bits of leaf, remains of the panicle, inflorescences, flowers and pedicels are used to feed sheep (Jacobsen et al. 2003).

4 Traits Contributing to Climate-Resiliency of Quinoa

4.1 Tolerance to Heat Stress

The ability of plants preserve high stomatal conductance at high temperature, inducing transpirational heat dissipation, is known as heat-tolerant plants (Kaushal et al. 2016). Hinojosa et al. (2019) demonstrated that quinoa could adapt to high day/night temperatures (40/24 °C) by showing high plasticity in terms of several growth parameters, such as net photosynthesis, chlorophyll fluorescence, leaf greenness index, seed size and seed set. They also stated that at high temperature, both quinoa genotypes exhibited high stomatal conductance. Yamori et al. (2014) stated that when plants exhibit considerable plasticity in a certain parameter, they tend to have great plasticity in other parameters; this is known as "syndrome of temperature acclimation". Consequently, quinoa plants under high temperature can still keep evaporative cooling during heat stress under sufficient water supply (Becker et al. 2017).

Yang et al. (2016) reported that the rise in temperature from (18/8 °C) to (25/20 °C) increased the plant height by 100% in the quinoa cv. "Titicaca". Hinojosa et al. (2019) indicated that although high temperature (40/24 °C) decreased pollen viability, it did not have a concomitant effect on seed yield. Their results also showed that growth parameters were not adversely affected when quinoa plants were exposed to high temperature, with the exception of pollen viability. Furthermore, (Maliro et al. 2017) indicated that quinoa genotypes grown in field with an average temperature of 20 °C had higher plant height than the plants of the same genotypes grown under average temperature of 15 °C in another location. Furthermore, exposing

Red Head, Salcedo and Cherry Vanilla quinoa cultivars to high- temperature treatment (35/29 °C) increased its stem mass. In addition, seed yield of the variety "Cherry Vanilla" decreased by 30%, whereas the variety "Salcedo" increased to 70% when the temperature was increased from 20/14 °C to 35/29 °C (Bunce 2017). In controlled experiments conditions, day/night temperatures of 25/20 °C stimulated quinoa growth (Yang et al. 2016).

4.2 *Tolerance to Salinity Stress*

The respond of quinoa plants to salinity stress occurs in two phases. In the first phase, a rapid response to the increase in external osmotic pressure, which starts immediately after the salt concentration around the roots increases to threshold levels, and that decreases new shoot growth. In the second phase, a slower response due to the accumulation of Na + in leaves (salt accumulation to toxic concentrations) and increases senescence of older leaves (Munns and Tester 2008). The tolerance of quinoa to salinity is a heritable trait with a polygenic character linked to a complex genetic basis that can be used as an efficient criterion for selection of salt resistant populations (Flowers and Colmer 2008). At the primary stages of seed germination, quinoa's tolerance to high salinity is based upon alterations in the levels of primary metabolites and enzyme activity (Adolf et al. 2013).

Kaya and Yazar (2016) indicated that quinoa irrigated with saline water of 10 and 30 dS/m resulted in 7 and 18% reduction in the yield, respectively in comparison to irrigation with fresh water. Razzaghi et al. (2011) reported that the seed yield of quinoa cultivar Titicaca was reduced under salinity up to 20 dS/m of the irrigation water, but when exposed to salinity levels higher than 20 dS/m, the plants adapted to it without yield components reduction. Similarly, Riccardi et al. (2014) reported that irrigation of quinoa plants with saline water (ECe = 22 dS/m) did not affect seeds quality.

In Egypt, several experiments were done to test its ability to tolerating salinity stress. Quinoa was cultivated in a greenhouse by Algosaibi et al. (2015) and irrigated it with four salinity levels (1.25, 4, 8 and 16 dS/m). They reported that the highest values of seed yield were recorded with 4 dS/m salinity level. Furthermore, 16 dS/m salinity level recorded the lowest seed yield of quinoa. Mahmoud (2017) conducted experiments in clay soil with salinity level equal to 3.37 dS/m. The plants were irrigated with three levels of saline water, namely 0.65, 10 and 20 dS/m. The results indicated that quinoa seed yield was negatively influenced by increasing water salinity, where about 12–21% reductions in seed yield were recorded in the plots irrigated with water having 10 dS/m. Furthermore, 40–45% reductions in seed yield of quinoa were recorded in the plots irrigated with water having 20 dS/m. Eisa et al. (2014) indicated that cultivation of quinoa under high saline soil conditions (ECe = 27 dS/m) resulted in 30% yield reduction compared to soil with low salinity level (ECe = 1.9 dS/m). On the other hand, Eisa et al. (2017) reported that quinoa seed yield was reduced under soil EC value equal to 17.9 dS/m, however seed quality was not highly affected. Eisa et al. (2017) tested the response of quinoa plants to high salinity under the Egyptian conditions. They concluded that quinoa seed yield was reduced as a result of increasing soil salinity up to 17.9 dS/m. Their results also indicated that there was a high ash content in seeds under saline conditions due to the increase of Na as well as K, P and Fe concentrations. By contrast, soil salinity led to significant decrease in Ca and Zn contents in seeds. Furthermore, Na was mainly accumulated in the pericarp followed by embryo tissues. They concluded that increasing most of essential minerals, especially Fe, in quinoa seeds produced under high saline conditions gave quinoa a distinctive value for human consumption. Eisa et al. (2018) tested the effect of two planting densities for quinoa, namely 56,000 plants/ha as low density and 167,000 plants/ha as high density in marginal soil of El-Fayoum, where Ec was 5.3 dS/m in irrigation water and 3.8 dS/m in soil paste. Their results indicated that a higher yield was obtained when the higher planting density was used. However, this high plant density was associated with significant reduction in seed quality in terms of protein content.

4.3 Tolerance to Water Stress

Water scarcity is a severe problem affecting plant growth and yield in the world. Under drought, many species respond by increasing the proportion of assimilates diverted to root growth with the concomitant root/shoot ratio increase (Sharp and Davies 1989). A decrease in both transpiration and CO_2 assimilation rates were observed under water deficit as a result of low cell turgidity and stomatal closure of the leaves (Waseem et al. 2011). Water stress affects cell turgidity and stomatal aperture of leaves resulting in decreases in both transpiration rates and CO_2 assimilation (Waseem et al. 2011). Under reduced rate of CO_2 assimilation and water stress, leaf metabolism is impaired. Furthermore, Albert et al. (2011) reported that the equilibrium between photochemical activity at photosystem II and electron requirement for photosynthesis is affected by water stress, generating an over excitation on photosynthetic system and photoinhibitory damages of photosystem II reaction centers. Adaptation to water stress has been associated with cell osmotic adjustment, which is accomplished by accumulation of different compounds, such as soluble sugars, proline, glycine betaine, polyols, and other organic compounds (Chai et al. 2001).

Soil moisture plays an important role in determining the time and rate of quinoa seed germination and seedling growth (Gonzalez et al. 2009). Quinoa develops unique resistance mechanisms to water stress, allowing the plant to adapt to harsh conditions in arid and semi-arid regions (Jacobsen et al. 2003). Moreover, using deficit irrigation with fresh water irrigate of quinoa resulted in reduction in yield, where application of 50 and 75% of full irrigation resulted in 6 and 9% reduction in the yield of quinoa (Kaya and Yazar 2016). Razzaghi et al. (2012) attained no significant differences between yield of quinoa cultivar Titicaca under full and deficit irrigated treatments in a clay-loam soil. Pulvento et al. (2012) showed the ability of the same cultivar to preserve yield under drought and soil salinity at a level of 16 dS/m in an open-field experiment conducted in South Italy in clay-loam soil. Similarly, Cocozza et al. (2013) recorded good resistance of this cultivar to water and salt stresses through stomatal responses and strict dependencies between relative water content and water potential components that caused osmotic adjustments, which is important for the maintenance of leaf turgor favorable to plant growth. Geerts (2008) showed that deficit irrigation is a valuable option for yield stabilization of quinoa plant in various areas of Bolivian Altiplano, where intra-seasonal dry periods are of great occurrence.

In Egypt, Algosaibi et al. (2017) indicated that reduction of irrigation water by 50% reduced the quinoa seeds by only 14%.

5 Quinoa Cultivation in Egypt

The productivity of traditional crops in the Egyptian cropping system is low when cultivated in marginal lands as a result of soil low fertility, high soil salinity and harsh weather conditions. In these environmental conditions, quinoa is considered as a good candidate to be cultivated there. Integrating quinoa as new crops in the Egyptian cropping pattern could play an important role in reducing wheat flour production-consumption gap. Because quinoa has a potential to grow as an alternative crop in marginal areas or with poor soil quality, it could contribute to increasing national agricultural production.

Bazile et al. (2016) reported that the FAO organization was actively involved in promoting and evaluating the cultivation of quinoa in 26 countries outside the Andean region with the aim to strengthen food and nutrition, where Egypt was one of these countries. Field evaluations were conducted in two seasons and in two sites using 13 genotypes. A higher value than the weighted average yield for all countries was obtained with 1.89 ton/ha at the first site and 2.35 ton/ha at the second Egyptian site.

A series of studies have been conducted on quinoa in Egypt to expectantly lead to its inclusion to innovate the cropping system of desert areas. Quinoa seemed to be very pertinent to grow in the Egyptian desert and new reclaimed sandy soil to combat degradation in these soils. Shams (2011) reported that thirteen varieties and strains of quinoa (from Europe and Peru) were grown in sandy soils of North Sinai in Egypt under two land preparation treatments, namely with or without tillage. The results indicated that the European varieties out-yielded the Peruvian varieties (1.93 ton/ha vs. 1.71 ton/ha, respectively). Furthermore, cultivation of quinoa under no tillage resulted in a higher yield, compared to cultivation with tillage (1.59 ton/ha vs. 1.49 ton/ha, respectively). Shams (2011) tested the effect of four sowing dates of quinoa seeds (15th of November, 15th of December, 1st of February and 15th of February) and two inter-row spacing, i.e. 15 and 20 cm apart in sandy soil. His results indicated that the highest yield was obtained when 15 cm inter-row spacing was used. Moreover, the yield of quinoa was reduced when planting date was delayed to February, where yield reduction was 63%. Shams (2012) conducted field trials in the sandy soils of Egypt North Nile Delta to determine the adequate nitrogen fertilizer rates to obtain higher yields of quinoa. The results showed that fertilizing quinoa with 360 kg N/ha increased plant height to reach 52.7 cm, increased grain yield per plant to reach 10.1 g/plant, increased grain yield to 1.20 ton/ha and increased biological yield to 2.79 ton/ha. Abdel-Rheem et al. (2014) indicated that the suitable sowing date for quinoa in Middle Egypt is 25th of November, where the highest yield was obtained. Furthermore, irrigation application when 60% of soil moisture was depleted (irrigation every 40 days) resulted in the highest yield. They also reported that water consumptive use for quinoa was between 215 and 226 mm.

According to Gomaa (2013), the highest values of quinoa yield in calcareous soil in Egypt were recorded with application of 240 kg/ha ammonium nitrate in combination with bio-fertilizer (Nitrobin), in addition to application with 120 kg/ha calcium super phosphate in combination with bio-fertilizer (Phosphorin). She also indicated that the above treatment increased crude protein and mineral elements (phosphorus, potassium and calcium) in seeds. Shams (2018) conducted field trials at Ismailia, Egypt to evaluate nine quinoa genotypes under arid environment of sandy soil and determine its agronomic potential and chemical composition. The results showed that all the studied quinoa genotypes were successful in sandy soils with good grain yield under Egyptian conditions. Season length of the studied genotypes ranged between 115 to 160 days in the field. The highest yield was obtained from Amarilla Sacaca quinoa genotype with a season length of 160 days. The highest protein content in quinoa grains was recorded from QS17-2 accession, while the lowest value was recorded by Blanca de Junin variety. Moreover, Salcedo INIA variety had the lowest saponins content in quinoa grains while QS16 accession recorded the highest content.

As stated by Mahmoud (2017), sowing quinoa in North Egypt in the beginning of winter season (last quarter of November where maximum air temperature was 18.8 °C and minimum temperature was 8.2 °C) resulted in higher plant height, biomass and seed yields, compared to the sowing at second quarters of December and January. The optimum growth was associated with relative humidity of 68.8% and cumulative daylight hours of 1819 h and cumulative sunshine hours of 977 h. Symptoms of powdery mildew disease were observed on the quinoa leaves 53 days after sowing and progressed in appearance. The high percentages of relative humidity could be the major cause of spreading the disease in quinoa.

6 The Advantage of Using Millets as a Food Crop

Millets is the sixth most important global cereal crop grown in the semi-arid regions of sub-Saharan Africa and the Indian subcontinent (Haussmann et al. 2012). It is particularly adapted to regions of low moisture and soils with low fertility. The short life-cycle of millets assists in escaping from stress as they require 12–14 weeks to complete their life-cycle (seed to seed), whereas rice and wheat requires a maximum of 20–24 weeks. It is the hardiest cereal crop that can be grown in a vast range of environmental harsh conditions, for instance, environments with high mean temperatures and frequent droughts and/or with poor soil fertility (Sehgal et al. 2015). Their deep root systems and short life cycles make it possible for millet to grow in places where the mean annual precipitation is as low as 300 mm (Panaud 2006). The unique chemical and physical properties of millet make it quite resistant to pests and diseases and also require little input during growth. The exceptional tolerance of millets toward diverse abiotic stresses including heat, salinity and drought makes them a tractable system to study their stress-responsive traits at the cellular, molecular and physiological levels (Muthamilarasan et al. 2016). These attributes make it a next-generation crop possess climate-resilient traits (Bandyopadhyay et al. 2017).

7 Nutritional Advantages of Pearl Millet

Millets are grouped into two subfamilies, namely Panicoideae, which includes pearl millet, foxtail millets, Japanese barnyard millet and Indian millet, and to Chloridoideae, which includes finger millet and tef, and eight genera (Tadele 2016). Millets are one of the world's six major cereal commodities, and it is consumed by one-third of the world's population. It hold great promise for food security and nutrition amid ever-increasing agricultural costs, climate change and increasing population (Kamatar et al. 2015). They are nutritionally superior to other major cereals as they are rich in dietary fibers, resistant starches, vitamins, essential amino acids, storage proteins and other bioactive compounds (Amadou et al. 2013). Millet-base food serves as the main source of energy, protein, vitamins and minerals, niacin, B6 and folacin, as well as essential amino acids such as isoleucine, leucine, phenylalanine, threonine and nitriloside compounds that play a role in inhibiting the development of cancer cells, and reduce the risk of heart disease (Kamatar et al. 2015). It is also rich in in vitamin B, folic acid, phosphorus, iron and potassium. (Gupta et al. 2011). Pearl millet seeds contained 11.8 g protein, 4.8 g fat, 2.2 g ash, 2.3 g crude fiber, 67.0 g carbohydrate, and energy amount of 363 kcal, which are lower than its values in quinoa. In addition, it contained 42 mg Ca and 11 mg Fe (Hulse et al. 1980; United States National Research Council).

In addition, millets are gluten-free, easy to digest and suitable for special diets, such as those associated to wheat intolerance, stomach ulcers or high cholesterol levels (Kamatar et al. 2015). Although there is little research in this area, it appears that millets are a great source of antioxidants and might have anticarcinogenic properties (Dykes and Rooney 2006). The flour made from the seeds of millet is used for making cakes and bread. The grains are also used as feed for birds, poultry, and ruminants (Kaume 2006).

8 Traits Contributing to Climate-Resiliency of Millets

8.1 Tolerance to Heat Stress

Millet is a warm weather crop with greater resistance to drought conditions, compared to other cereal crops grown in warm dry regions. Being a C4 photosynthesis, millets provides improved growth and ecological enactment in warm temperatures, enhanced flexible allocation patterns of biomass and reduced hydraulic conductivity per unit leaf area (Sage and Zhu 2011). It is highly tolerant to heat because its root branching and flowering speed increase along with the growth of new leaves at certain high temperature levels (Singh et al. 2003). It has the potential to germinate over varying temperatures from 15 to 40 °C (Yadav et al. 2006). Garcia-Huidobro et al. (1985) revealed that germination rate and final germination percentage were not reduced following short exposure of high temperature, namely 45 °C. The relative growth rate and net assimilation rate are high in pearl millet (Ashraf and Hafeez 2004). However, exposure to more than 40 °C decreases the germination percentage and rate of emergence. Optimum soil temperature for germination is 23–30 °C, and the crop complete germination within 2–4 days under favorable conditions (Sher et al. 2019). Furthermore, optimum temperature range for maximum net photosynthesis is between 35 and 40 °C. This crop is very susceptible to low temperature at seedling and reproductive stages.

8.2 Tolerance to Salinity Stress

Millets are rated to be fairly tolerant to salinity and have superiority over wheat, corn, and sorghum under dry and saline conditions in semi-arid land (Krishnamurthy et al. 2007). Yakubu et al. (2010) reported that, plant height, shoot and root dry weights of five pearl millet varieties were significantly decreased with increasing soil salinization. Furthermore, Radhouane (2013) applied three levels of saline water irrigation, namely T1 (1 g NaCl/l as control), T2 (4 g NaCl/l) and T3 (7 g NaCl/l) to pearl millet plants. The results showed that plant height of T3 was 18% significantly lower than T1 and 13% than T2. Reiad et al. (2014) indicated that applying saline water irrigation of 4500 ppm did not reduce forage yield of pearl millet and its attributes.

8.3 Tolerance to Water Stress

Millet is grown in tropical and subtropical climates owing to its low water requirement compared with other cereals. Being a C4 crop, millets are characterized by having high water use efficiency, which is ~1.5 to fourfold higher than C3 photosynthesis

(Aubry et al. 2011). It is one of the most drought tolerant crop plants because of its short development stages, capacity to grow speedily under favorable conditions, rapid growth after stress, and effective control of water loss and heat tolerance (Bidinger et al. 1987). This is due to being a C4 plants, where it can prevent photorespiration and, as a consequence, efficiently utilize the scarce moisture present in the semi-arid regions. Since C4 plants are able to close their stomata for long periods, they can significantly reduce moisture loss through the leaves (Tadele 2016).

Pearl millet is highly adapted to drought. Sivakumar (1992) stated that pearl millet matches its phenology to the mean distribution of the rainfall where precipitation is limited and erratic. In this case, the development of the main panicle coincided with an increasing period of rain, thus reducing the risks associated with drought events occurring prior to or at the beginning of flowering. It also adjusts flowering according to the pattern of rainfall (Bidinger et al. 2007) and reduces water extraction before flowering and increase the extraction after flowering (Vadez et al. 2013). Seed number and biomass of pearl millet (Aparna et al. 2014) and seed yield (Vadez et al. 2013) were found to be unaffected under drought. Furthermore, Ajithkumar and Panneerselvam (2014) observed an increase in root length in little millet under drought stress. Furthermore, millet changes growth rate and plant architecture in response to water availability to plants to maximize yield in drought prone areas (Winkel and Do 1992).

The critical stages of pearl millet for irrigation are flowering and grain development. However, as most of the crops are grown in rain-fed area in the world, the irrigation cannot be applied in those areas, and the whole crop completes its life cycle on the soil moisture available from rainfall (Sher et al. 2019). Under a certain degree of drought stress, pearl millet has a higher resistance to drought than other species of millet (finger millet, Job's tears, barnyard millet, common millet, and foxtail millet) and maintains an unchanged morphology, particularly with respect to the leaf area and shoot fresh and dry weight (Kusaka et al. 2010). Li and Brutnell (2011) indicated that foxtail millet requires just 257 g of water to produce a dry biomass of 1 g, whereas maize and wheat require 470 and 510 g of water, respectively, which implies high water use efficiency in millet species than other cereal crops. van der Weerd et al. (2001) showed the dynamics of membrane permeability for water in pearl millet is higher than that of maize for achieving better water status during osmotic stress.

9 Pearl Millet Cultivation in Egypt

Unfortunately, pearl millet is cultivated in Egypt as a forage crop not for human consumption. In this context, Shahin et al. (2013) indicated that green forage yield, as well as dry forage yield were significantly affected by the interaction between nitrogen fertilization rates and cutting height, where the highest yield was detected when 144–180 kg N/ha was applied and cutting height was 20 cm. Ziki et al. (2019) indicated that millet forage cutting at 63 days after sowing followed by 42 and 35 days interval for the following cuts, with application of 108 kg N/ha per after each cut obtained better growth and green forage. Salama et al. (2020) compared between four genotypes and local variety of pearl millet. They found that among these new genotypes, IP13150 maintained the desirable balance between both productivity and quality, in addition to its high dry matter yield, and highest crude protein, with the lowest fiber fractions. This was reflected on its organic matter digestibility, high gas production, short-chain fatty acid production, microbial protein, and the highest energy values among all the genotypes. Whereas, the local variety was inferior regarding all the tested quality attributes.

Recently, there are some studies in Egypt promoting mixing pearl millet flour with wheat flour in bread making (Ali and Halim 2013; Abd El-Kader 2016) because wheat protein is lacking in some of essential amino acids such as lysine, threonine and valine (Czubaszek and Karolini-Skaradziñska 2005).

10 The Advantage of Using Sorghum as a Food Crop

Sorghum is a multipurpose crop belonging to the Poaceae family, which are C4 carbon cycle plants with high photosynthetic efficiency and productivity (Ayers and Westcot 1985). Sorghum is one of the most important cereal crops in semiarid area because of its tolerance to drought and high temperatures compared with other cereal crops (Abraha et al. 2015). Additionally, this crop has tremendous potentiality to grow under adverse condition by minimum input and (Khaton et al. 2016), where and it can be grown and can maintain remarkable yield potential in environments that are normally too extreme for other C4 plants (Promkhambut et al. 2011). It is capable to grow and produce yield at low rainfall area, where it is cultivated in areas considered too dry and hot for other cereals, because of its tolerance to drought and heat stress (Kenga et al. 2006).

11 Nutritional Advantages of Sorghum

Sorghum can grow as fodder and for human consumption in areas relay under water stress (Khalili et al. 2008). Sorghum has a high concentration of potassium and starch, it is less acidifying and is easily absorbed and tolerated by the sick and diabetics, adults and even children. Sorghum a gluten-free and can substitute wheat in bread making (Ismail and El-Nakhlawy 2019). Sorghum seeds contained 10.4 g protein, 3.1 g fat, 1.6 g ash, 2.0 g crude fiber, 70.7 g carbohydrate, and energy amount of 329 kcal. In addition, it contained 25 mg Ca and 5.4 mg Fe (Hulse et al. 1980; United States National Research Council). It is rich in antioxidants, which are believed to help lower the risk of cancer, diabetes, heart disease and some neurological diseases (Ismail and El-Nakhlawy 2019).

When freshly chopped, sorghum can be used to feed cows, goats, sheep, and even chickens because it has the same energy levels as maize and other cereals. Sorghum produces much more forage than maize (Miller and Stroup 2004).

12 Traits Contributing to Climate-Resiliency of Sorghum

12.1 Tolerance to Heat Stress

Sorghum is known for its ability to survive under extreme hot and dry environments, has highly conserved flowering behavior, with 100% flowering in a day completed within the first hour from dawn., which helps maintain pollen viability under heat stress conditions (Chiluwal et al. 2019). Sorghum pollen is highly sensitive to heat stress but a large genetic variation for pollen viability exists that can help in reducing heat stress-induced crop yield losses in sorghum (Sunoj et al. 2017). Sorghum cultivars require soil temperatures above 21 °C, while maize can germinate at much lower soil temperatures, namely 13 °C, which effectively ensures that sorghum is exposed to higher temperatures during much of its growth cycle, not just the flowering and grain-filling phases (Shroyer et al. 1998).The optimum range of air temperature for sorghum during the vegetative period is 26–34 °C (Hammer et al. 1993); and during the reproductive period is 25–28 °C (Prasad et al. 2008).

12.2 Tolerance to Salinity Stress

Sorghum is believed to be tolerate to soil and water salinity up to 6.8 and 4.5 dS/m of electrical conductivity, respectively (Ayers and Westcot 1985). In a saline environment, sorghum showed a certain ability to exclude Na (Yang et al. 2018) and limit Na transport from the roots to the leaves by unloading Na from the xylem into the roots (Shakeri and Emam 2017). Additionally, sorghum can compartmentalize Na into the cell vacuoles as an osmolyte to adjust osmosis at the cellular level and thereby compensate the potential drop in the growing medium (De Lacerda et al. 2005). Selective uptake and translocation of K and Ca versus Na were identified as further mechanism for salt tolerance in sorghum (Shakeri et al. 2020). Furthermore, sorghum has additional strategies to maintain cellular osmotic pressure, defend plant metabolism against reactive oxygen species, and protect the assembly of photosystems under salinity through the accumulation of osmoprotectants such as proline (Nxele et al. 2017) and sugars (Yang et al. 2020), the increase in chlorophylls and carotenoids levels (Yilmaz et al. 2020), and the enhanced antioxidant enzymatic activity (Hussain et al. 2019). The salt-resistant genotypes can successfully cope with osmotic and ionic stresses caused by the excess of NaCl; they can effectively reduce the oxidative damage and are able to detoxify the harmful metabolites (Zhu 2001).

12.3 Tolerance to Water Stress

Sorghum is known by its ability to tolerate water shortage due to its roots system, which can extend into the deep soil layers, as well as being highly productive C4 photosynthetic plant (Lynch et al. 2014). Sorghum developed two important strategies for adaptation to water deficit. The primary way is the tolerance of water potential decrease, while the second mechanism is the escape from water stress due to deep and extensive root formation (Tari et al. 2013). Drought-tolerant genotypes can be characterized by high epicuticular wax deposition on the leaf surface, which increases leaf reflectance in the visible and near-infrared radiation and contributes to decreased transpiration, which is another important function of the leaf epicuticular wax is to reduce the cuticular conductance to water vapor and thus to increase water-use efficiency (Surwenshi et al. 2010).

Furthermore, drought tolerant/resistant sorghum varieties are highly water use efficient with an extensive root system for water absorption and a thick waxy cuticle that retards excessive water loss (Schittenhelm and Schroetter 2014). Different sorghum varieties also accumulate varying amounts of osmolytes (Devnarain et al. 2016) for osmotic adjustment during periods of drought stress. Sorghum is known for maintaining its yield output even under drought conditions, and this requires protection of the photosynthetic machinery from drought-induced osmotic and oxidative damage known to disrupt photosynthetic pigments (Clarke and Durley 1981). Blum (2005) observed an increase in the root/shoot ratio as an adaptive feature for enhanced soil water uptake in order to maintain favorable cellular hydration during drought stress. Furthermore, he also reported that root length and depth may increase in dry conditions even with a reduced total root mass.

Khatab et al. (2017) evaluated water stress tolerance of some sorghum genotypes, where their results confirmed that some genotypes exhibited highly resistance traits for drought. Similar results were obtained for other sorghum cultivar by Jabereldar et al. (2017) and Tawfik and El-Mouhamady (2019). Kothari et al. (2020) found that enhancing drought tolerance by increasing root density at different soil depths resulted in higher grain sorghum yield. Sarshad et al. (2021) indicated that sorghum is tolerant to water stress during grain dough stage and the end of grain filling stage.

13 Sorghum Cultivation in Egypt

Grain sorghum is one of the most important grain crops for vast number of people in Upper Egypt, where it is used in bread making. Azzazy et al. (2001) indicated that the highest yield of sorghum was obtained under using drip system for irrigation and the application of 240 kg N/ha. El-Aref et al. (2005) compered between growing two sorghum hybrids under drip or surface irrigation systems and they found that the highest grain yield was obtained from hybrid Shandaweel–2 when plants received 300 kg N/ha under irrigation using drip irrigation system. The appropriate sorghum plant density, applied irrigation water, and nitrogen fertilizer for highest yield were determined by Salem (2015). His results showed that plant density of 84,000 plant/ha, irrigation with 16.8 mm/ha/day, and application of nitrogen fertilizer of 240 kg N/ha achieved the maximum values of grain, straw and biological yields. Ismail and El-Nakhlawy (2019) indicated that application of humic acid with 80% of the required irrigation to sorghum cause low yield losses, compared to application of 100% of the required irrigation.

14 Conclusion

Quinoa, millets and sorghum are three important climate-resilient crops that can contribute in building resilience to climate and mitigate the effect of climate variability and change on food security. These three crops are tolerant to heat, salinity and water stresses, which allow it to be grown in harsh environments not suitable for other cereals. They can also play a significant role in the livelihood of the population of developing world due to their enormous nutritious contents. However, these crops have not been sufficiently studied to realize its maximum capability to resist climate variability and change and produce the highest productivity.

References

Abd El-Kader MH (2016) Egyptian seven seeds bread. Middle East J Appl Sci 06(02):403–410

Abdellatif ASA (2018) Chemical and technological evaluation of quinoa (Chenopodium Quinoa Willd) cultivated in Egypt. Acta Sci Nutr Health 2:115–125

Abdel-Rheem HA, Abdrabbo A, Aboukheira AA, Eisa NMA (2014) Response of quinoa crop for water stress and planting date in the middle Egypt. J Egypt Acad Soc Environ Dev 15:49–66

Abraha T, Githiri S, Kasili R, Araia W, Nyende A (2015) Genetic variation among Sorghum (Sorghum bicolor L. Moench) landraces from Eritrea under post-flowering drought stress conditions. Am J Plant Sci 6:1410–1424

Adolf VI, Jacobsen SE, Shabala S (2013) Salt tolerance mechanisms in quinoa (Chenopodium quinoa Willd.). Environ Exp Bot 92:43–54

Ajithkumar IP, Panneerselvam R (2014) ROS scavenging system, osmotic maintenance, pigment and growth status of Panicum sumatrense Roth under drought stress. Cell Biochem Biophys 68:587–595. https://doi.org/10.1007/s12013-013-9746-x

Albert KR, Mikkelsen TN, Michelsen A, RoPoulsen H, van der Linden L (2011) Interactive effects of drought, elevated CO_2 and warming on photosynthetic capacity and photosystem performance in temperate heath plants. J Plant Physiol 168:1550–1561

Algosaibi AM, El-Garawany MM, Badran AE, Almadini AM (2015) Effect of irrigation water salinity on the growth of quinoa plant seedlings. J Agric Sci 8:208–213

Algosaibi AM, Badran AE, Almadini AM, El-Garawany MM (2017) The effect of irrigation intervals on the growth and yield of quinoa crop and its components. J Agric Sci 9:182–191

Ali MA, Halim MA (2013) Bread made from wheat flour and white sesame flour in presences of natural improver. Pakistan J Nutr 12(4):353–355

Amadou I, Gounga ME, Le GW (2013) Millets: nutritional composition, some health benefits and processing—a review. Emir. J. Food Agric. 25:501–508. https://doi.org/10.9755/ejfa.v25i7.12045

Aparna K et al (2014) Seed number and 100-seed weight of pearl millet [Pennisetum glaucum L.) respond differently to low soil moisture in genotypes contrasting for drought tolerance. J Agron Crop Sci 200(2):119–131

Arendt EK, Zannini E (2013) Cereal grains for the food and beverage industries. Food science, technology and nutrition. Woodhead Publishing 248:409–438

Ashraf M, Hafeez M (2004) Thermotolerance of pearl millet and maize at early growth stages: growth and nutrient relations. Biol Plant 48(1):81–86

Aubry S, Brown NJ, Hibberd JM (2011) The role of proteins in C3 plants prior to their recruitment into the C4 pathway. J Exp Bot 62:3049–3059. https://doi.org/10.1093/jxb/err012

Ayers RS, Westcot DW (1985) Salinity problems. In: Water quality for agriculture; irrigation and drainage, Paper 29. FAO, Rome, Italy, pp 13–56. ISBN 9251022631

Azzazy NB, El-Shafai AMA, Abazaied AA, Osman HE (2001) Effect of irrigation methods and nitrogen fertilization on yield and quality of sweet sorghum in Upper Egypt. Zeitschrift Für Bewässerungswirtschaft 36(2):193–201

Bandyopadhyay T, Muthamilarasan M, Prasad M (2017) Millets for next generation climate-smart agriculture. Front Plant Sci 8:1266. https://doi.org/10.3389/fpls.2017.01266

Bazile D, Bertero HD, Nieto C (eds) (2015) State of the art report on quinoa around the world in 2013. FAO; CIRAD, Rome. http://www.fao.org/3/a-i4042e.pdf

Bazile D, Pulvento C, Verniau A, Al-Nusairi MS, Ba D, Breidy J, Hassan L, Mohammed MI, Mambetov O, Otambekova M, Sepahv NA, Shams A, Souici D, Padulosi MK (2016) Worldwide evaluations of quinoa: preliminary results from post international year of quinoa FAO Projects in Nine Countries. Front Plant Sci 7:850–855

Becker VI, Goessling JW, Duarte B, Caçador I, Liu F, Rosenqvist E, Jacobsen S-E (2017) Combined effects of soil salinity and high temperature on photosynthesis and growth of quinoa plants (Chenopodium quinoa). Funct Plant Biol 44:665–678. https://doi.org/10.1071/FP16370.

Bidinger FR, Mahalakshmi V, Rao GDP (1987) Assessment of drought resistance in pearl millet [Pennisetum americanum (L.) Leeke]. I. Factors affecting yield under stress. Aust J Agric Res 38:37–48

Bidinger FR, Nepolean T, Hash CT, Yadav RS, Howarth CJ (2007) Quantitative trait loci for grain yield in pearl millet under variable post flowering moisture conditions. Crop Sci 47:969–980. https://doi.org/10.2135/cropsci2006.07.0465

Blum A (2005) Drought resistance, water-use efficiency, and yield potential-are they compatible, dissonant, or mutually exclusive? Aust J Agric Res 56:1159–1168

Bunce JA (2017) Variation in yield responses to elevated CO_2 and a brief high temperature treatment in quinoa. Plants 6:26. https://doi.org/10.3390/plants6030026

Chai CL, Li SH, Xu YC (2001) Carbohydrate metabolism in peach leaves during water stress and after stress relief. Plant Physiol Commun 37:495–498

Chiluwal A, Bheemanahalli R, Kanaganahalli V, Boyle D, Perumal R, Pokharel M, Oumarou H, Jagadish SVK (2019) Deterioration of ovary plays a key role in heat stress-induced spikelet sterility in sorghum. Plant Cell Environ 43:448–462

Clarke JM, Durley RC (1981) The responses of plants to drought stress. In: Simpson GM (ed) Water stress on plants. Praeger Publishers New York, pp 89–139

Cocozza C, Pulvento C, Lavini A, Riccardi M, d'Andria R, Tognetti R (2013) Effects of increasing salinity stress and decreasing water availability on eco-physiological traits of quinoa grown in a Mediterranean-type agroecosystem. J Agron Crop Sci 199:229–240

Czubaszek A, Karolini-Skaradzińska Z (2005) Effects of wheat flour supplementation with oat products on dough and bread quality. Pol J Food Nutr Sci 14/55(3):281–286

De Lacerda CF, Cambraia J, Oliva MA, Ruiz HA (2005) Changes in growth and in solute concentrations in sorghum leaves and roots during salt stress recovery. Environ Exp Bot 54:69–76

Devnarain N, Crampton BG, Chikwamba R, Becker JVM, O'Kennedy MM (2016) Physiological responses of selected African sorghum landraces to progressive water stress and re-watering. S Afr J Bot 103:61–69. https://doi.org/10.1016/j.sajb.2015.09.008

Dhankher OP, Foyer CH (2018) Climate resilient crops for improving global food security and safety. Plant Cell Environ 41:877–884

Dida MM, Devos KM (2006) Finger millet. Cereals and Millets. Springer, Berlin, pp 333–343

Dykes L, Rooney LW (2006) Sorghum and millet phenols and antioxidants. J Cereal Sci 44:236–251

Eisa S, Abdel-Ati A, Ebrahim M, Eid M, Abd El-Samad E, Hussin S, El-Bordeny N, Ali S, El-Naggar A (2014) Chenopodium quinoa as a new non-traditional crop in Egypt. In: Proceeding of "Bridging the gap between increasing knowledge and decreasing resources" conference, Tropentag, September 17–19, Prague, Czech Republic

Eisa SE, Eid MA, Abd El-Samad EH, Hussin SA, Abdel-Ati AA, El-Bordeny NE, Ali SH, Hanan MAA, Lotfy ME, Masoud AM, El-Naggar AM, Ebrahim M (2017) Chenopodium quinoa Willd. A new cash crop halophyte for saline regions of Egypt. Aust J Crop Sci 1:343–351

Eisa SS, Abd El-Samad EH, Hussin SA, Ali EA, Ebrahim M, González JA, Ordano M, Erazzú LE, El-Bordeny NE, Abdel-Ati AA (2018) Quinoa in Egypt—plant density effects on seed yield and nutritional quality in marginal regions. Middle East J Appl Sci 8:515–522

El-Aref KhAO, Abdel-Mawly SE, Abo-Elhamd AS (2005) Improving yield and water use efficiencies of two sorghum cultivars irrigated by surface and drip irrigation systems and fertilized by nitrogen. Ass Univ Bull Environ Res 8(2):67–73

Escuredo O, Martín MIG, Moncada GW, Fischer S, Hierro JMH (2014) Amino acid profile of the quinoa (Chenopodium quinoa Willd.) using near infrared spectroscopy and chemometric techniques. J Cereal Sci 60:67–74

FAO (2013) Launch of the international year of quinoa: UN celebrates Andean super food. http://www.fao.org/quinoa-2013/pressroom/news/detail/en/

Flowers TJ, Colmer TD (2008) Salinity tolerance in halophytes. New Phytol 179:945–963

Francis G, Kerem Z, Makkar HP, Becker K (2002) The biological action of saponins in animal systems: a review. Br J Nutr 88:587–605

Fuentes-Bazan S, Mansion G, Borsch T (2012) Towards a species level tree of the globally diverse genus Chenopodium (Chenopodiaceae). Mol Phylogenet Evol 62:359–374

Garcia M, Raes D, Jacobsen SE, Michel T (2007) Agro-climatic constraints for rain fed in the Bolivian Altiplano. J Arid Environ 71:109–121

Garcia M, Condori B, Castillom CD (2015) Agroecological and agronomic cultural practices of Quinoa in South America. In: Murphy K, Matanguihan J (ed) Quinoa: improvement and sustainable production, 1st edn. Wiley

Geerts S (2008) Deficit irrigation strategies via crop water productivity modeling: field research of quinoa in the Bolivian Altiplano. Dissertationes de Agricultura, Faculty of Bio-Science Engineering, K.U. Leuven, Belgium

Gollin D, Morris M, Byerlee D (2005) Technology adoption in intensive post-green revolution systems. Am J Agric Econ 87:1310–1316

Gomaa EF (2013) Effect of nitrogen, phosphorus and biofertilizers on quinoa plant (Chenopodium quinoa). J Appl Sci Res 9:5210–5222

Gonzalez JA, Gallardo M, Mirna HM, Rosa M, Prado FE (2009) Physiological responses of quinoa (Chenopodium quinoa Willd.) to drought and waterlogging stresses: dry matter partitioning. Bot Stud 50:35–42

Gracia-Huidobro J, Monteith JL, Squire GR (1985) Time, temperature and germination of pearl millet (Pennisetum typhoides S & H). III. Inhibition of germination by short exposure to high temperature. J Exptl Bot 36:338–343

Gupta N, Gupta AK, Singh NK, Kumar A (2011) Differential expression of PBF Dof transcription in different tissues of three finger millet genotypes differing in seed protein content and color. Plant Mol Biol Rep 29:69–76. https://doi.org/10.1007/s11105-010-0208-y

Hammer GL, Carberry PS, Muchow RC (1993) Modeling genotypic and environmental control of leaf area dynamics in grain sorghum. Whole plant level. Field Crops Res 33:293–310. https://doi.org/10.1016/0378-4290(93)90087-4

Haussmann BIG, Fred Rattunde H, Weltzien-Rattunde E, Traoré PSC, vom Brocke K, Parzies HK (2012) Breeding strategies for adaptation of pearl millet and sorghum to climate variability and Change in West Africa. J Agron Crop Sci 198:327–339

Hinojosa L, Matanguihan JB, Murphy KM. (2019) Effect of high temperature on pollen morphology, plant growth and seed yield in quinoa (Chenopodium quinoa Willd.). J Agro Crop Sci 205:33–45. https://doi.org/10.1111/jac.12302.

Hirose Y, Fujita T, Ishii T, Ueno N (2010) Antioxidative properties and flavonoid composition of Chenopodium quinoa seeds cultivated in Japan. Food Chem 119:1300–1306. https://doi.org/10.1093/jxb/err179

Hulse JH, Liang EM, Pearson OE (1980) Sorghum and the millets: their composition and nutritive value. Academic Press, New York

Hussain S, Shaukat M, Ashraf M, Zhu C, Jin Q, Zhang J (2019) Salinity stress in arid and semi-arid climates: effects and management in field crops. In: Hussain S (ed) Climate change and agriculture. InTech Open: Rijeka, Croatia, pp 655–201

Ismail SM, El-Nakhlawy FS (2019) Optimizing grain sorghum (sorghum bicolor l.) productivity under full irrigation and stress using humic acid in arid regions. Assiut J Agric Sci 50(2):272–288

Jabereldar AA, El Naim AM, Abdalla AA, Dagash YM (2017) Effect of water stress on yield and water use efficiency of sorghum (Sorghum bicolor L. Moench) in semi-arid environment Int J Agric For 7(1):1–6. https://doi.org/10.5923/j.ijaf.20170701.01

Jacobsen SE, Mujica A, Jensen CR (2003) The resistance of quinoa (Chenopodium quinoa Willd) to adverse abiotic factors. Food Rev Intl 19:99–109

Jacobsen SE, Jensen CR, Liu F (2012) Improving crop production in the arid Mediterranean climate. Field Crops Res 128:34–47

Kamatar MY, Brunda SM, Sanjeevsingh HH, Sowmya GG, Hundekar R (2015) Nutritional composition of seventy five, elite germplasm of foxtail millet (Setaria italica). Int J Eng Res 4(4):1–6

Kaume RN (2006) Panicum miliaceum L. In: Brink M, Belay G (eds) Plant resources of Tropical Africa 1. Cereals and pulses. PROTA Foundation, Wageningen, Netherlands, pp 122–126

Kaushal N, Bhandari K, Siddique KHM, Nayyar H (2016) Food crops face rising temperatures: an overview of responses, adaptive mechanisms, and approaches to improve heat tolerance. Cogent Food Agric 2:1134380. https://doi.org/10.1080/23311932.2015.1134380

Kaya CI, Yazar A (2016) Saltmed model performance for quinoa irrigated with fresh and saline water in a Mediterranean environment. Irrig Drain 65:29–37

Kenga R, Tenkouano A, Gupta SC, Alabi SO (2006) Genetic and phenotypic association between yield components in hybrid sorghum (Sorghum bicolor (L.) Moench) populations. Euphytica 150:319–326. https://doi.org/10.1007/s10681-006-9108-5

Khalili I, Akbari N, Chaichi MR (2008) Limited irrigation and phosphorus fertilizer effects on yield and yield components of grain sorghum (Sorghum bicolor L. var. Kimia). Am-Eurasian J Agric Environ Sci 3:697–702

Khatab IA, El-Mouhamady AA, Abdel-Rahman HM, Farid MA, El-Demardash IS (2017) Agro-morphological and molecular characterization of sorghum (Sorghum vulgare L.) for water stress tolerance. Int J Curr Res Biosci Plant Biol 4:37–55

Khaton MA, Sagar A, Tajkia JE, Islam MS, Mahmud MS, Hossain AKMZ (2016) Effect of moisture stress on morphological and yield attributes of four sorghum varieties. Progress Agric 27:265–271

Kothari K, Ale S, Bordovsky JP, Porter DO, Munster CL, Hoogenboom G (2020) Potential benefits of genotype-based adaptation strategies for grain sorghum production in the Texas high plains under climate change. Eur J Agron 117:126037. https://doi.org/10.1016/j.eja.2020.126037A

Krishnamurthy L, Serraj R, Rai KN, Hash CT, Dakheel AJ (2007) Identification of pearl millet [Pennisetum glaucum (L.) R. Br.] lines tolerant to soil salinity. Euphytica 158(1–2):179–188

Kusaka M, Ohta M, Fujimura T (2010) Contribution of inorganic components to osmotic adjustment and leaf folding for drought tolerance in pearl millet. Physiol Plant 125(4):474–489

Li P, Brutnell TP (2011) Setaria viridis and Setaria italica, model genetic systems for the panicoid grasses. J Exp Bot 62:3031–3037. https://doi.org/10.1093/jxb/err096

Lin BB (2011) Resilience in agriculture through crop diversification: adaptive management for environmental change. Bio Sci 61:183–193

Lynch JP, Chimungu JG, Brown KM (2014) Root anatomical phenes associated with water acquisition from drying soil: targets for crop improvement. J Exp Bot 65(21):6155–6166

Mahmoud AH (2017) Production of Quinoa (Chenopodium quinoa) in the marginal environments of South Mediterranean Region: Nile Delta, Egypt. Egypt J Soil Sci 57:329–337

Maliro MFA, Guwela VF, Nyaika J, Murphy KM (2017) Preliminary studies of the performance of quinoa (Chenopodium quinoa Willd.) Genotypes under irrigated and rainfed conditions of central Malawi. Front Plant Sci 8:227. https://doi.org/10.3389/fpls.2017.00352

Maradini-Filho AM, Pirozi MR, Borges JTS, Santana HMP, Chaves JBP, Reis-Coimbra JSD (2017) Quinoa: nutritional, functional and anti-nutritional aspects. Crit Rev Food Sc. Nutr 57:1618–1630

Massawe F, Mayes S, Cheng A (2016) Crop diversity an unexploited treasure trove for food security. Trends Plant Sci 21:365–368

Miller FR, Stroup JA (2004) Growth and management of sorghum for forage production. http://alfalfa.ucdavis.edu/+symposium/proceedings/2004/04-149.pdf

Molina-Montenegro MA (2013) Quinoa biodiversity and sustainability for food security under climate change. A Review. Agron Sustain Dev 34:349–359

Morita N, Hirata C, Park SH, Mitsunaga T (2001) Quinoa flour as a new food stuff for improving dough and bread. J Appl Glyco Sci 48:263–270

Munns R, Tester M (2008) Mechanisms of salinity tolerance. Annu Rev Plant Biol 59:651–681

Muthamilarasan M, Dhaka A, Yadav R, Prasad M (2016) Exploration of millet models for developing nutrient rich graminaceous crops. Plant Sci 242:89–97. https://doi.org/10.1016/j.plantsci.2015.08.023

Nxele X, Klein A, Ndimba BK (2017) Drought and salinity stress alters ROS accumulation, water retention, and osmolyte content in sorghum plants. S Afr J Bot 108:261–266

Panaud O (2006) Foxtail millet. Cereals and millets. Springer, Berlin, pp 325–332

Prasad PVV, Pisipati SR, Mutava RN, Tuinstra MR (2008) Sensitivity of grain sorghum to high temperature stress during reproductive development. Crop Sci 48:1911–1917. https://doi.org/10.1016/j.jplph.2009.11.007

Promkhambut A, Polthanee A, Akkasaeng C, Younger A (2011) Growth, yield and aerenchyma formation of sweet and multipurpose sorghum (Sorghum bicolor L. Moench) as affected by flooding at different growth stages. Aust J Crop Sci 5(8):954

Pulvento C, Riccardi M, Lavini A, Iafelice G, Marconi E, d'Andria R (2012) Yield and quality characteristics of quinoa grown in open field under different saline and non-saline irrigation regimes. J Agron Crop Sci 198:254–263

Radhouane L (2013) Agronomic and physiological responses of pearl millet ecotype (Pennisetumglaucum (L.) R. Br.) to saline irrigation. Emir J Food Agric 25(2):109–116

Razzaghi F, Ahmadi SH, Adolf SH, Jensen CR, Jacobsen SE, Andersen MN (2011) Water relations and transpiration of quinoa (Chenopodium quinoa Willd.) under salinity and soil drying. J Agr Crop Sci 197:348–360

Razzaghi F, Plauborg F, Jacobsen SE, Jensen CR, Andersen MN (2012) Effect of nitrogen and water availability of three soil types on yield, radiation use efficiency and evapotranspiration in field-grown quinoa. Agric Water Manag 109:20–29

Reiad MSh, Maha MA, Hamada MSh, Abd El-Maaboud, Khalil MH (2014) Forage growth and productivity of pearl millet as affected by soil mulching, planting date under salinity conditions. Egypt J Agron 36(1):75–94

Riccardi M, Pulvento C, Lavini A, d'Andria R, Jacob sen SE (2014) Growth and ionic content of quinoa under saline irrigation. J Agro Crop Sci 200:246–260

Rojas W, Pinto M, Alanoca C, Gomez Pando L, LeonLobos P, Alercia A (2015) Quinoa genetic resources and ex situ conservation. In: Bazile D, Bertero HD, Nieto C (eds) State of the art report on Quinoa around the world in 2013. FAO & CIRAD, Roma, pp 56–82.R

Romero S, Shahriari S (2011) Quinoa's global success creates quandary at home. The New York Times, 19 March 2011. http://www.nytimes.com/2011/03/20/world/americas/20bolivia.html?_r=2&hp=&pag&

Ruiz KB, Aloisi I, Duca SD, Canelo V, Torrigiani P, Silva H, Biondi S (2016) Salares versus coastal eco-types of quinoa: salinity responses in Chilean landraces from contrasting habitats. Plant Physiol Biochem 101:1–13

Sage RF, Zhu XG (2011) Exploiting the engine of C4 photosynthesis. J Exp Bot 62:2989–3000

Salama HSA, Shaalan AM, Nasser MEA (2020) Forage performance of pearl millet (Pennisetum glaucum [L.] R. Br.) in arid regions: yield and quality assessment of new genotypes on different sowing dates. Chilean J Agric Res 80(4):572–584

Salem EMM (2015) Response of grain sorghum (sorghum bicolor, l. Monech) to irrigation, nitrogen and plant density under new valley conditions, Egypt. Egypt J Desert Res 65(1):11–30

Sarshad, Talel D, Torabi M, Rafiel F, Nejatkhan P (2021) Morphological and biochemical responses of sorghum bicolor (L) Moench under drought stress. Springer Nat Appl Sci 3(18). https://doi.org/10.1007/s42452-020-03977-4

Schittenhelm S, Schroetter S (2014) Comparison of drought tolerance of maize, sweet sorghum and sorghum-sudangrass hybrids. J Agron Crop Sci 200:46–53

Sehgal D, Skot L, Singh R, Srivastava RK, Das SP, Taunk J et al (2015) Exploring potential of pearl millet germplasm association panel for association mapping of drought tolerance traits. PLoS ONE 10(5):e0122165. https://doi.org/10.1371/journal.pone.0122165

Shahin MG, Abdrabou RTh, Abdelmoemn WR, Hamada MM (2013) Response of growth and forage yield of pearl millet (Pennisetum galucum) to nitrogen fertilization rates and cutting height. Ann Agric Sci 58(2):153–162

Shakeri E, Emam Y (2017) Selectable traits in sorghum genotypes for tolerance to salinity stress. J Agric Sci Technol 19:1319–1332

Shakeri E, Emam Y, Pessarakli M, Tabatabaei SA (2020) Biochemical traits associated with growing sorghum genotypes with saline water in the field. J Plant Nutr 43:1136–1153

Shams AS (2011) Combat degradation in rain fed areas by introducing new drought tolerant crops in Egypt. Int J Water Resour Arid Environ 1:318–325

Shams AS (2012) Response of quinoa to nitrogen fertilizer rates under sandy soil conditions. In: Proceeding of the 13th international conference in agronomy, Benha, Egypt, 9–10 September, pp 195–205

Shams AS (2018) Preliminary evaluation of new quinoa genotypes under sandy soil conditions in Egypt. Agric Sci 9:1444–1456

Sharp RE, Davies WJ (1989) Regulation of growth and development of plants growing with a restricted supply of water. In: Jones HG, Flowers TJ, Jones MB (eds) Plants under stress. Cambridge University Press, England, pp 71–93

Sher A, Nawaz A, Sarfraz M, Ijaz M, Ul-Allah S, Sattar A, Hussain S, Ahmad Sh (2019) Advanced production technologies of millets. In: Agronomic crops. Springer Nature Singapore Pte Ltd

Shroyer J, Kok H, Fjell D (1998) Seedbed preparation and planting practices. Grain sorghum production handbook (Kansas State Univ, Manhattan, KS). https://www.bookstore.ksre.ksu.edu/pubs/C687.pdf

Singh R, Singh D, Tyagi P (2003) Effect of Azotobacter, farmyard manure and nitrogen fertilization on productivity of pearl millet hybrids (Pennisetum glaucum (l) r. br) in semi-arid tropical environment. Arch Agron Soil Sci 49(1):21–4

Sivakumar MVK (1992) Empirical-analysis of dry spells for agricultural applications in West Africa. J Clim 5(5):532–539

Smith SM (2012) Food crisis in the Horn of Africa: progress report, July 2011–July 2012 (Oxfam International)

Soliman AH, Abbas MS, Abol-Ella MF, Eassawy MT, Mohamed RH (2019) Towards bridging wheat gap in Egypt by using cassava, quinoa and guar as supplements for the production of balady bread. J Food Meas Charact 13:1873–1883

Sunoj VSJ, Somayanda IM, Chiluwal A, Perumal R, Prasad PVV, Jagadish SVK (2017) Resilience of pollen and post-flowering response in diverse sorghum genotypes exposed to heat stress under field conditions. Crop Sci 57:1658–1669

Surwenshi A, Chimmad VP, Jalageri BR, Kumar V, Ganapathi M, Nakul HT (2010) Characterization of sorghum genotypes for physiological parameters and yield under receding soil moisture conditions. Res J Agric Sci 1:242–244

Tadele Z (2016) Drought adaptation in millets. In: Shanker A, Shanker C (eds) Abiotic and biotic stress in plants: recent advances and future perspectives. InTech Open, pp 639–662

Tari I, Laskay G, Takács Z, Poór P (2013) Response of sorghum to abiotic stresses: a review. J Agron Crop Sci 199:264–274

Tawfik RS, El-Mouhamady ABA (2019) Molecular genetic studies on abiotic stress resistance in sorghum entries through using half diallel analysis and inter-simple sequence repeat (ISSR) markers. Bull Natl Res Cent 43:117

Vadez V, Kholová J, Yadav RS et al (2013) Small temporal differences in water uptake among varieties of pearl millet (Pennisetum glaucum (L.) R. Br.) are critical for grain yield under terminal drought. Plant Soil 371:447–462. https://doi.org/10.1007/s11104-013-1706-0

van der Weerd L, Claessens MM, Ruttink T, Vergeldt FJ, Schaafsma TJ, Van As H (2001) Quantitative NMR microscopy of osmotic stress responses in maize and pearl millet. J Exp Bot 52:2333–2343. https://doi.org/10.1093/jexbot/52.365.2333

Vilche C, Gely M, Santalla E (2003) Physical properties of quinoa seeds. Biosys Eng 86:59–65

Waseem M, Ali A, Tahir M, Nadeem M, Ayub M, Tanveer A (2011) Mechanisms of drought tolerance in plant and its management through different methods. Cont J Agric Sci 5:10–25

Winkel T, Do F (1992) Caracteres morphologiques etphysiologiques de resistance du mil millet[Pennisetum glaucum (L.) R. Br.] a la secheresse. L'agronomie Tropicale 46:339–350

Wu G, Peterson AJ, Morris CF, Murphy KM (2016) Quinoa seed quality response to sodium chloride and sodium sulfate salinity. Front Plant Sci 7:1–8

Yadav AK, Narwal MS, Singh B (2006) Field screening technique for heat effect on seedlings of pearl millet. In: Proceedings of the national seminar on transgenic crops in Indian agriculture: status, risk and acceptance, pp 109–112, CCS HAU, Hisar, India

Yakubu H, Ngala AL, Dugje IY (2010) Screening of millet (Pennisetum glaucum L.) varieties for salt tolerance in semi-arid soil of northern Nigeria. World J Agric Sci 6(4):374–380

Yamori W, Hikosaka K, Way DA (2014) Temperature response of photosynthesis in C3, C4, and CAM plants: temperature acclimation and temperature adaptation. Photosynth Res 119:101–117. https://doi.org/10.1007/s11120-013-9874-6

Yang A, Akhtar SS, Amjad M, Iqbal S, Jacobsen S-E (2016) Growth and physiological responses of quinoa to drought and temperature stress. J Agron Crop Sci 202:445–453. https://doi.org/10.1111/jac.12167

Yang Z, Zheng H, Wei X, Song J, Wang B, Sui N (2018) Transcriptome analysis of sweet Sorghum inbred lines differing in salt tolerance provides novel insights into salt exclusion by roots. Plant Soil 430:423–439

Yang Z, Li JL, Liu LN, Xie Q, Sui N (2020) Photosynthetic regulation under salt stress and salt-tolerance mechanism of sweet sorghum. Front Plant Sci 10:1–12

Yilmaz S, Temizgül R, Yürürdurmaz C, Kaplan M (2020) Oxidant and antioxidant enzyme response of redbine sweet sorghum under NaCl salinity stress. Bioagro 32:31–38

Zhu J-K (2001) Plant salt tolerance. Trends Plant Sci 6:66–72

Ziki SJL, Zeidan EMI, El-Banna AYA, Omar AEA (2019) Growth and forage yield of pearl millet as influenced by cutting date and nitrogen fertilization. Zagazig J Agric Res 46(5):1351–1361

Chapter 7
Assessment of Climate Variability on Wheat Productivity in Egypt

Samiha Ouda

Abstract In this chapter, multiple regression analysis was employed to find out which weather element have contributed more in wheat productivity and how strong is the contribution. Two sets of data were used in the analysis. The first set composed of seasonal values of weather data for 10 years (2010–2019), namely solar radiation (SR, MJ/m^2/day), maximum temperature (Mx, °C), minimum temperatures (Mn, °C) and relative humidity (RH, %) existed in each agro-climatic zone and the corresponded average values of wheat productivity in the zone. It was used to assess the spatial-variability of wheat zone-specific productivity from one agro-climatic zone to another, as a result of the studied weather elements in each zone. The second data set contained the above mentioned seasonal values collected for 34 years (1986–2019) in each agro-climatic zone and the corresponding values of national wheat productivity. It was used to determine the temporal-variability of national wheat productivity from year to year, as affected by the variability in each weather elements overall the agro-climatic zones. The results of first data set showed that seasonal Mx was the major contributor in wheat zone-specific productivity in a negative trend, in all agro-climatic zones, except the first agro-climatic zone. Seasonal Mn also negatively contributed in wheat zone-specific productivity in the second, third, fourth agro-climatic zones. Whereas, seasonal RH only contributed positively in the first agro-climatic zone and negatively in the second agro-climatic zone. Additionally, seasonal SR did not have any contribution in wheat zone-specific productivity. Furthermore, the analysis of the second data set revealed that there was higher variability existed in the studied climate elements in the past 10–12 years, compared to the rest of the data set. It also revealed that SR have a positive significant contribution in national wheat productivity in Zone1 and Zone4. Mx and Mn had a negative significant contribution in all zones. Furthermore, RH have a negative significant contribution Zone5 only. Thus, these results implied that the contribution of seasonal weather elements on zonal level, is different than its contribution of on the national level. Furthermore, these results ascertained that the Egyptian wheat cultivars are resilient to climate variability

S. Ouda (✉)
Water Requirements and Field Irrigation Research Department, Soils, Water and Environment Research Institute; Agricultural Research Center, Giza, Egypt

S. Ouda and A. E.-H. Zohry, *Climate-Smart Agriculture*,
https://doi.org/10.1007/978-3-030-93111-7_7

and possess yield stability traits because it have been exposed to unfavorable weather conditions caused variability in its production, with insignificant reduction.

Keywords Weather elements · Multiple linear regression analysis · Coefficient of determination (R^2) · Root mean square error (RMSE) · Willmott index of agreement (*d-stat*)

1 Introduction

The World Meteorological Organization (2012) stated that weather reflects short-term conditions of the atmosphere while climate is the average daily weather for an extended period of time at a certain location over many years (usually at least 30 years). The uncertainty associated with weather patterns is one of the greatest contributors to the gap presented between potential and actual yield (Zhang et al. 2013), where it was estimated by 20% in many cases (Lobell et al. 2009). Two weather elements directly affect plant growth and photosynthesis, namely temperature and solar radiation (Wassmann et al. 2009). Global warming and environmental pollution have increased the average global surface temperature and decreased the solar radiation, which in turn, have increased the frequency of environmental stresses, such as heat stress and solar dimming, both adversely affecting the growth and yield of crops (Ambavaram et al. 2014).

Wheat is an important crop worldwide and a major contributor in food security. Grain yield in wheat is determined mainly by environmental factors (Villegas et al. 2015). In Inner Mongolia, Zhao et al. (2017) found that the influence of the average maximum temperature on the key phenophases of spring wheat was greater than the average minimum temperature, followed by the relative humidity and solar radiation. Phenological adjustment, or the optimization of the duration of the different developmental phases, has been one of the most useful strategies available to adapt wheat to harsh or/and highly erratic environmental conditions (Gouache et al. 2012). Time to anthesis is considered a primary trait determining wheat adaptation to a particular set of growing conditions. Variability in time to anthesis can be used to fine-tune growth and development patterns to the most prevailing environmental conditions in any particular environment (Blum 2011). Royo et al. (2010) reported that the percentage of yield variability explained by the environment could reach 98%, where mean daily maximum temperature from emergence to heading could account for 59% of yield variations.

In Egypt, wheat is the major crop in the daily diet, where it is the main crop used in bread making. The cultivated area of wheat is the highest among the winter crops, where it competes with Egyptian clover and most recently with sugar beet. Most of the wheat cultivated area lies in the Nile Delta with about 57% of the total area, and smaller areas in Middle and Upper Egypt with about 18 and 17%, respectively and the rest is cultivated in the new reclaimed areas (Zohry and Ouda 2020). Wheat yields in Egypt increased by 5.8-fold between 1961 and 2017 due to varietal improvement and the use of improved managements techniques, such as the raised bed cultivation under surface irrigation, optimum sowing date, land levelling, optimum fertilizers, and implementing wheat intercropping systems with other suitable crops on raised beds (Abdelmageed et al. 2019).

Climate variability have been observed in Egypt and it negatively affected the growing crops. It has been reported that the heat wave occurred in February in 2010 negatively affected national wheat production (Khalil and Hassanein 2016). The rise in both maximum and minimum temperature occurred and it was higher by 3.5 and 2.6 °C, respectively than the values recorded in 2009. That heat weave resulted in 15% losses in the national wheat production, compared to the production in 2009. Wheat plants were exposed to other heat weaves in Egypt since 2010, however no such high decline in wheat productivity has been reported. This is due to the advances in wheat breeding and the release of new cultivars with high yield stability under abiotic stress, as well as to the matching of local climatic conditions with plant growth and development process of these cultivars (Abdelmageed et al. 2019). Deng et al. (2015) indicated that matching local climatic conditions with plant growth and development could results in obtaining high yield.

Egypt does not produce enough wheat to feed its population, thus it fulfill their needs by importation from different countries, which make Egypt the largest wheat importer globally. Wheat self-sufficiency declines from 51% in 2013 to 43% in 2017 (Zohry and Ouda 2018). This decline is a result of the rise in wheat consumption by the population, which is increases annually. Consequently, the policy of the government of Egypt has been to implement breeding programs for wheat to produce new tolerant cultivars to abiotic stresses and to attain yield stability. The Egyptian wheat cultivars are suited to be grown throughout Egypt, in both old lands as well as in the new reclaimed areas. Additionally, technical recommendations to attain high wheat yield have been also provided to the farmers.

In this chapter, multiple linear regression analysis was used to test whether the Egyptian wheat cultivars are resilient to climate variability and possess yield stability traits under the experienced adverse weather conditions prevailed lately.

2 The Egyptian Wheat Cultivars

Huge improvements in wheat production have been reported by the Ministry of Agriculture and Land Reclamation (2019) in Egypt. Abdelmageed et al. (2019) indicated that the achieved increase in wheat yield was done through the breeding and release of varieties with shorter plant height to resist lodging, has long spikes and large number of tillers. The breeding was done also for improving nitrogen use efficiency, rusts resistance, as well as increase tolerance to water, heat and salinity stresses. Abou Gabal and Tabl (2014) identified two tolerant wheat cultivars to heat stress during germination stage, namely Sakha93 and Gemmeiza9. Furthermore, Abdelmageed et al. (2019) listed 50 wheat cultivars (bread and durum) released between 1947 and 2012. These cultivars are tolerant to different biotic and abiotic stresses.

Table 1 List of selected tolerant wheat cultivars to biotic and abiotic stresses

Cultivar	Region	Characteristics	Average yield (ton/ha)	Cultivated area (ha)[a]
Sakha93	Lower Egypt	Rusts resistant and heat tolerant	6.5	6,766
Giza168	All governorates	Water deficit, heat tolerant and rusts resistant	6.5	120,184
Gemmeiza9	Lower Egypt, Giza, and Fayoum (Middle Egypt)	Resistant to rusts	6.5	45,402
Gemmeiza12	Lower Egypt, Toshka/El-Owinate (South west Egypt)	Resistant to rusts, and heat tolerant	8.2	11,747
Misr1	All governorates Toshka/El-Owinate	Resistant to rusts, water deficit, and heat tolerant	8.6	295,983
Benisuif1	Middle and Upper Egypt, Toshka/El-Owinate	Resistant to yellow rust, and heat tolerant	6.3	50,260
Benisuif5	Middle and Upper Egypt, Toshka/El-Owinate	Heat tolerant	7.4	102,596

Source Abdelmageed et al. (2019)
[a] *Source* Ministry of Agriculture and Land Reclamation (2019), Egypt in the growing season of 2018/19

For the subject of our study, we presented some of these cultivars in Table 1. In the wheat growing season of 2018/19, the cultivated area of these cultivars was 48% of the total wheat cultivated area in this season.

3 Effect of Weather Elements on Wheat Growth and Yield

3.1 Temperature

Temperature regulation is foundational in biological systems, where any chemical reaction rates are a function of the tissue temperature and the energy required to initiate the reaction (Wolfenden 2003). Heat stress is expressed as the rise in air temperature beyond a threshold level for a period sufficient to cause injury or irremediable damage in plants in general (Teixeira et al. 2013). In wheat, heat stress affects various plant processes leading to morphophysiological alterations in plants, hindering the development processes and eventually resulting into great yield loss (Grant et al. 2011). High temperature decreases metabolic activities (Farooq et al.

2009), and production of oxidative reactive species (Wang et al. 2011). Howarth (2005) indicated that high temperature have direct and indirect injuries on wheat plants, namely direct injuries include protein denaturation and aggregation, and increased fluidity of membrane lipids. Whereas, indirect or slower heat injuries include inactivation of enzymes in chloroplast and mitochondria, inhibition of protein synthesis, protein degradation and loss of membrane integrity.

The primary effect of heat stress is the impediment of seed germination causing poor stand establishment in wheat (Hossain et al. 2013) and that negatively affects plant meristems (Kosova et al. 2011). Almeselmani et al. (2009) observed that high temperature (35/25 °C) imposed after tillering showed a significant reduction in water potential in wheat and reduces plant growth by promoting leaf senescence and abscission (Kosova et al. 2011). Heat stress in wheat during anthesis causes reduction in pollen tube development, and increases pollen mortality (Oshino et al. 2011). Plants exposed to temperatures above 24 °C during reproductive stage significantly reduced grain yield and yield reduction continued with increasing duration of exposure to high temperature (Prasad and Djanaguiraman 2014). It also reduces the number of grains leading to lower harvest index in wheat (Lukac et al. 2011). Increase in temperature by 1–2 °C reduces seed mass by accelerating seed growth rate and shortening grain-filling periods in wheat (Nahar and Gretzmacher 2002). Heat stress speeds up the rate of seed filling by reducing the duration of this stage (Kaushal et al. 2016).

3.2 Solar Radiation

The increase of clouds and aerosols has reduced total solar radiation reaching the earth's surface (Ren et al. 2013). The site latitude is an important and integrative environmental driver, as it is associated with variations in temperature regimes, photoperiod and radiation intensity, all of which determine growth, development and productivity of a crop (Craufurd and Wheeler 2009).

Solar radiation is important element in crops growth, development and production through interception by the crop canopy (Chen et al. 2003). It plays an important role as regulator and controller of growth and development and influences assimilation of nutrients and dry matter distribution (Han et al. 2008). The principal process is the photosynthetic assimilation that makes synthesize vegetal components from water, CO_2 and the light energy possible (Lecoeur and Ney 2003). The solar radiation brings energy to the metabolic process of the plants, where it used in the evaporation process inside the different organs of the plants, and also in the transpiration through the stomas (Han et al. 2008). Ahmed et al. (2011) indicated that there is a direct relationship between solar radiation and wheat yield, where solar radiation is the major weather element in photosynthesis process.

3.3 Relative Humidity

The main climatic factors affecting disease severity are humidity, temperature, and wind (Lyon and Broders 2017). In Egypt, the disease that affected most of the Egyptian wheat cultivars, are rusts. It causes severe losses in grain yield reaches more than 23%, and in epidemic years, they may reach up to 50% (Draz et al. 2015). Stripe rust of wheat needs free moisture (high humidity, rainfall, or dew) for infection (Wiese 1987), and has an optimal temperature range around 10–15 °C (Hoggs et al. 1969). Draz et al. (2018) indicated that relative humidity was positively associated with stripe rust infection in wheat in two locations in Lower and Middle Egypt. Grain weight loss was strongly associated with the value of average coefficient of infection among cultivars for both locations, recording $r = 0.9044$ for Lower Egypt location and $r = 0.8866$ for Middle Egypt location.

4 Methodology of the Analysis

4.1 Collected Data

Weather data for the five agro-climatic zones of Egypt were collected from the following website: https://power.larc.nasa.gov. The five agro-climatic zones of Egypt was developed by Ouda and Noreldin (2017) using evapotranspiration values (ETo). Figure 1 presented the map of Egypt with its five agro-climatic zones.

Daily data was downloaded from the above site and monthly averages from November to April (wheat growing season) were calculated. The seasonal values of the selected weather elements were calculated for each agro-climatic zone. The latitude, longitude and the elevation above sea level of each agro-climatic zone are presented in Table 2.

Two sets of data were available and used in the analysis. The first set is a decade of weather data (2010–2019), composed of solar radiation (MJ/m^2/day), maximum and minimum temperatures (°C) and relative humidity (%) existed in each agro-climatic zone. In addition, the corresponded average wheat productivity values (ton/ha) in each zone were collected. Hence, this analysis determined the spatial-variability of wheat zone-specific productivity from one agro-climatic zone to another, as a result of the studied weather elements in each zone.

The second set is climate data contained the above mentioned seasonal values collected for 34 years from (1986–2019) in the five agro-zones of Egypt. Furthermore, the national wheat productivity in the above mentioned years were also collected. Thus, this analysis determined the temporal-variability of national wheat productivity from year to year, as affected by the variability in each weather elements overall the agro-climatic zones.

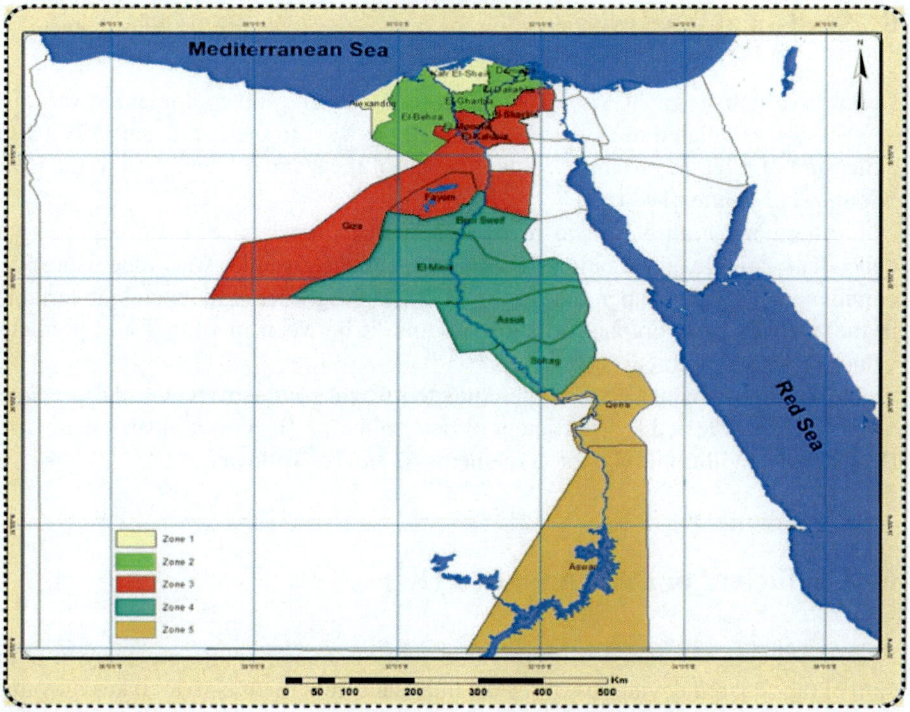

Fig. 1 Map of agro-climatic zones of Egypt. *Source* Ouda and Noreldin (2017)

Table 2 Latitude, longitude and elevation above sea level of agro-climatic zones of Egypt

Agro-climatic zones	Latitude (°)	Longitude (°)	Elevation above sea level (m)
First agro-climatic zone	31.70–31.07	29.00–30.57	7.00–20.00
Second agro-climatic zone	31.25–30.47	30.28–32.14	5.00–14.80
Third agro-climatic zone	30.36–29.18	31.01–30.51	13.00–30.00
Fourth agro-climatic zone	29.04–26.36	30.44–31.38	30.40–71.00
Fifth agro-climatic zone	26.10–24.02	32.43–32.53	72.60–108.30

5 Statistical Analysis

Descriptive statistics analysis for both data sets was done and coefficient of variation (CV%) was calculated to assess the variability existed in these data sets. CV (%) is a measure of relative variability. It is the ratio of the standard deviation to the mean (Gomez and Gomez 1984).

Furthermore, multiple linear regression analysis between wheat productivity as dependent variable and weather data as independent variables was done using 95% confidence interval (Draper and Smith 1987). The significant independent variables in the analysis were graphed to show its line fit between measured and predicted values of wheat productivity.

The goodness of fit between the collected wheat productivity and the predicted wheat yield was tested by coefficient of determination (R^2), root mean square error (RMSE) and Willmott index of agreements (*d-stat*) as followed:

6 Coefficient of Determination (R^2)

The R^2 value is calculated from the total sum of squares, more precisely, it is the sum of the squared deviations of the original data from the mean. R^2 takes on values between 0 and 1.0, a value of 1.0 indicates a perfect match, and value of 0 indicates no agreement at all. R^2 also shows how many points fall on the regression line (Jamieson et al. 1998).

$$R^2 = 1 - \frac{\sum_{i=1}^{n} (O_i - S_i)^2}{\sum_{i=1}^{n} \left(O_i - \overline{O}\right)^2} \tag{1}$$

where O_i and S_i represent the observed and predicted values, respectively, n represents the number of observed and predicted values used in comparison and $\overline{O}$ is the observed average.

7 Root Mean Square Error (RMSE)

It is the standard deviation of the residuals (prediction errors). Residuals are a measure of how far from the regression line data points are. It is a frequently used to measure of the difference between values predicted by a model and those actually observed from an experiment that is being modeled (Jamieson et al. 1998).

$$RMSE = \sqrt{\left(\frac{\sum_{i=1}^{n} (S_i - O_i)^2}{n}\right)}(2) \tag{2}$$

8 Willmott Index of Agreement (*d-stat*)

It is the standardized measure of the degree of model prediction error and varies between 0 and 1.0. A value of 1.0 indicates a perfect match, and value of 0 indicates no agreement at all (Willmott 1981).

$$d - stat = 1 - \frac{\sum_{i=1}^{n} (O_i - S_i)^2}{\sum_{i=1}^{n} \left[\left(|(S_i - \overline{O})| + |(O_i - \overline{O})| \right)^2 \right]} (3) \qquad (3)$$

9 Results and Discussion

9.1 Assessment of the Spatial-Variability of Wheat Zone-Specific Productivity and Weather

9.1.1 Spatial-Variability of Wheat Zone-Specific Productivity

As stated earlier, this analysis determined the spatial-variability of wheat zone-specific productivity from one agro-climatic zone to another, as a result of the studied weather elements in each zone.

Descriptive statistics of the seasonal values of wheat productivity in the agro-climatic zones of Egypt are presented in Table 3. The results indicated that the range between minimum and maximum wheat productivity within the zone was lower than 1.00 ton/ha in the first and second agro-climatic zones. Whereas, in the third and fourth agro-climatic zones, the range was lower than 1.50 ton/ha. But, in the fifth agro-climatic zone, that range increase to reach 3.22 ton/ha due to low minimum value of wheat productivity in that zone. The values of CV% followed that same trend of the value of range found in all zones, where the highest value was found in the fifth agro-climatic zone and the lowest value was found in the first agro-climatic zone.

Table 3 Descriptive statistics of wheat yield (ton/ha) in the agro-zones of Egypt from 2010 to 2019

Statistic	Zone 1	Zone 2	Zone 3	Zone 4	Zone 5
Mean (ton/ha)	6.38	6.76	7.20	6.97	6.36
Standard error	0.07	0.08	0.11	0.10	0.29
Sample variance	0.04	0.07	0.13	0.10	0.82
Range (ton/ha)	0.67	0.85	1.25	1.18	3.22
Minimum value (ton/ha)	5.88	6.29	6.26	6.26	3.86
Maximum value (ton/ha)	6.55	7.14	7.52	7.44	7.08
CV (%)	3.27	3.91	4.99	4.46	14.24

Table 4 Descriptive statistics of the seasonal weather elements in the first agro-climatic zone from 2010 to 2019

Statistic	SRAD	TMAX	TMIN	RH
Mean	14.71	21.93	14.54	63.06
Standard error	0.35	0.28	0.26	0.59
Standard deviation	1.10	0.87	0.82	1.87
Range	3.33	2.61	2.35	6.59
Minimum value	13.29	20.74	13.53	58.22
Maximum value	16.63	23.35	15.88	64.81
CV (%)	7.50	3.97	5.61	2.96

SRAD = solar radiation (MJ/m^2/day), TMAX = maximum temperature (ºC), TMIN = minimum temperature (ºC), and RH = relative humidity (%)

9.1.2 Spatial-Variability of the Weather Elements

The First Agro-climatic Zone

Descriptive statistics of the studied seasonal weather elements revealed that the highest range between minimum and maximum values was found for seasonal relative humidity and the lowest value was found for seasonal minimum temperature. Furthermore, the highest value of CV% was found for seasonal solar radiation and the lowest value was found for seasonal relative humidity (Table 4). These results implied the high variation in solar radiation values, which could be due to high cloud cover and/or air pollution existed in the first agro-climatic zone.

Multiple linear regression analysis showed a significant relationship ($P < 0.05$) between seasonal weather elements and wheat productivity in the first agro-climatic zone, where R^2 value was 0.6319, which means that 63.19% of the variation in wheat productivity can be explained by the studied weather elements. The goodness of fit analysis showed that RMSE value between measured and predicted wheat productivity was 0.1267 ton/ha and *d-stat* value was 0.8743.

Although higher variability in seasonal solar radiation existed in this zone, it did not have a significant effect on wheat productivity. Instead, the results showed that a significant positive relationships between seasonal relative humidity ($P < 0.05$) and wheat productivity were found in the first agro-climatic zone (Fig. 2). This positive relationship might be attributed to the amount of rain fall, which encourage farmers to reduce the applied irrigation water to wheat. It might be also attributed to cultivation of rusts tolerant wheat cultivars in that zone. Furthermore, there were insignificant relationships between the rest of the studied weather elements and wheat productivity.

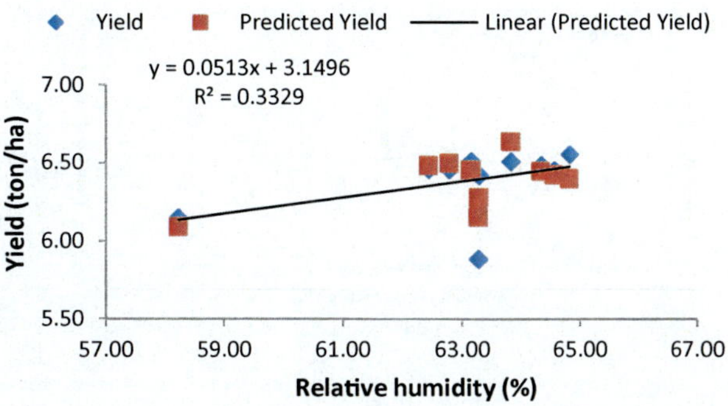

Fig. 2 Line fit between wheat productivity and seasonal relative humidity in the first agro-climatic zone

Table 5 Descriptive statistics of seasonal weather elements in the second agro-climatic zone from 2010 to 2019

Statistic	SRAD	TMAX	TMIN	RH
Mean	15.81	23.31	11.25	56.38
Standard error	0.35	0.38	0.25	1.44
Standard deviation	1.11	1.19	0.78	4.55
Range	3.94	3.25	2.12	14.08
Minimum value	13.28	21.85	10.23	48.79
Maximum value	17.22	25.10	12.36	62.88
CV (%)	6.99	5.11	6.92	8.06

SRAD = solar radiation (MJ/m^2/day), TMAX = maximum temperature (ºC), TMIN = minimum temperature (ºC), and RH = relative humidity (%)

The Second Agro-climatic Zone

The results in Table 5 indicated that, in the second agro-climatic zone, high variation in the seasonal relative humidity was found, manifested in the highest values of the range between minimum and maximum values of relative humidity, as well as highest value of CV%. This could be a result of rain fall amounts in the second agro-climatic zone, which is normally higher than the other zones. Whereas, the lowest values of both statistics were found for seasonal minimum temperature, which implied less variability.

Multiple linear regression analysis showed a significant relationship ($P < 0.05$) between the studied seasonal weather elements and wheat productivity in the second agro-climatic zone, where R^2 value was 0.9003, which means that 90.03% of the variation of wheat yield can be explained by the studied weather elements. Furthermore, the goodness of fit analysis indicated that RMSE value between measured and predicted wheat productivity was 0.1149 ton/ha and *d-stat* value was 0.9455.

A significant negative relationships between seasonal maximum temperature ($P < 0.05$) (Fig. 3a), and seasonal minimum temperature ($P < 0.05$) (Fig. 3b) with

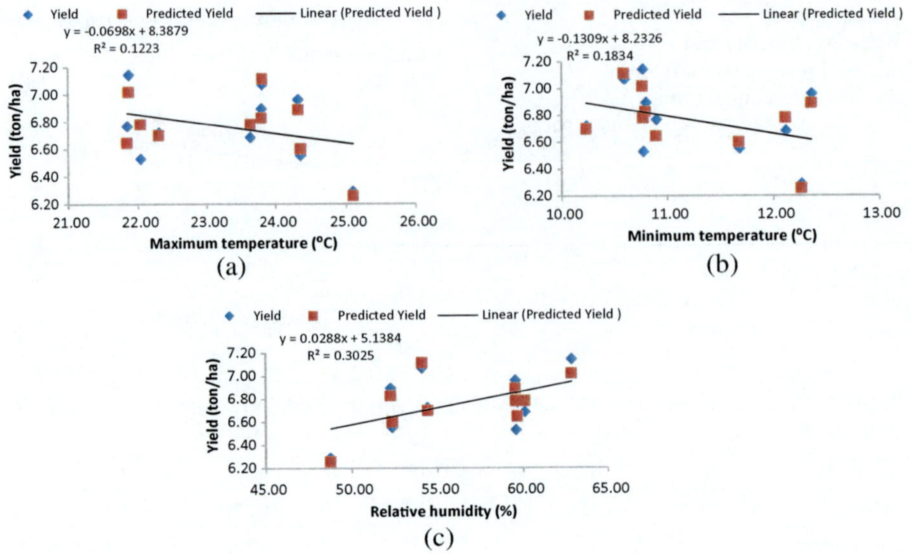

Fig. 3 Line fit between wheat productivity and seasonal maximum temperature (**a**), seasonal minimum temperature (**b**), and seasonal relative humidity (**c**) in the second agro-climatic zone

wheat productivity were found in the second agro-climatic zone. Moreover, a significant positive relationship between seasonal relative humidity ($P < 0.01$) and wheat productivity was found in that zone (Fig. 3c). Seasonal relative humidity displayed a higher value of CV% (Table 5).

The Third Agro-climatic Zone

With respect to the third agro-climatic zone, the analysis showed that the highest values of the range between minimum and maximum values was found for seasonal relative humidity and the highest value of CV% was found for minimum temperature. Whereas, the lowest value of range between minimum and maximum values was found for seasonal solar radiation and the lowest value for CV% was found for seasonal maximum temperature, respectively (Table 6).

A significant relationship ($P = 0.05$) resulted from multiple linear regression analysis between seasonal weather elements and wheat productivity in the third agro-climatic zone was found, where R^2 value was 0.6884. Furthermore, the goodness of fit analysis calculated low RMSE value between measured and predicted wheat productivity, namely 0.2908 ton/ha and high value of *d-stat* value, namely 0.8080.

Figure 4a showed high values of R^2 between seasonal solar radiation, seasonal maximum temperature (Fig. 4b), as well as seasonal minimum temperature (Fig. 4c) and wheat productivity, demonstrated negative relationships.

Table 6 Descriptive statistics of seasonal weather elements in the third agro-climatic zone from 2010 to 2019

Statistic	SRAD	TMAX	TMIN	RH
Mean	15.69	24.36	10.92	52.41
Standard error	0.25	0.37	0.32	1.16
Standard deviation	0.80	1.15	1.03	3.67
Range	2.39	3.68	3.34	13.25
Minimum value	14.65	22.70	9.46	46.49
Maximum value	17.04	26.37	12.80	59.74
CV (%)	5.12	4.74	9.40	7.01

SRAD = solar radiation (MJ/m^2/day), TMAX = maximum temperature (°C), TMIN = minimum temperature (°C), and RH = relative humidity (%)

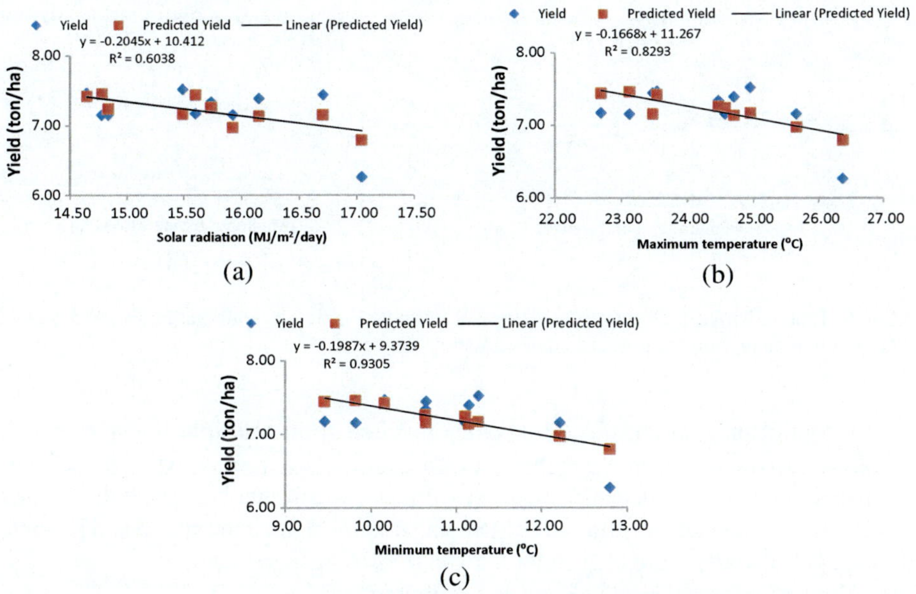

Fig. 4 Line fit between wheat productivity and seasonal solar radiation (**a**), seasonal maximum temperature (**b**), and minimum temperature (**c**) in the third agro-climatic zone

The Fourth Agro-climatic Zone

In the fourth agro-climatic zone (Table 7), the analysis showed that the highest values of range between minimum and maximum values and CV% were found for seasonal relative humidity. Whereas, the lowest value of the range between minimum and maximum values was found for seasonal solar radiation and the lowest CV% value was found for seasonal maximum temperature (Table 7).

Table 7 Descriptive statistics of seasonal weather elements in the fourth agro-climatic zone from 2010 to 2019

Statistic	SRAD	TMAX	TMIN	RH
Mean	17.29	24.68	10.64	43.56
Standard error	0.33	0.38	0.36	1.51
Standard deviation	1.03	1.21	1.13	4.76
Range	3.61	3.78	3.93	15.58
Minimum value	14.78	23.13	9.00	37.22
Maximum value	18.40	26.91	12.93	52.80
CV (%)	5.97	4.89	10.64	10.93

SRAD = solar radiation (MJ/m^2/day), TMAX = maximum temperature (°C), TMIN = minimum temperature (°C), and RH = relative humidity (%)

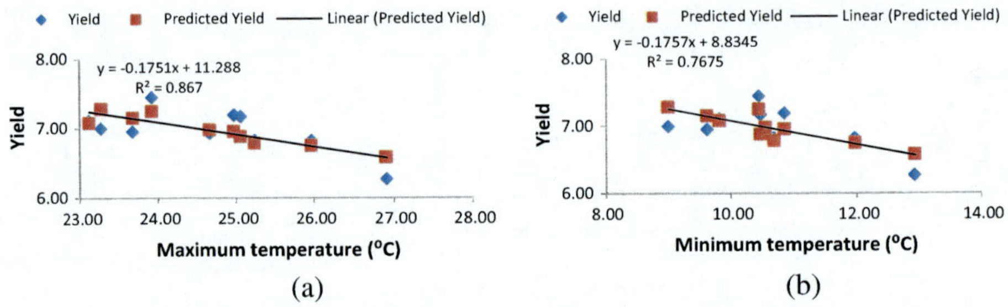

Fig. 5 Line fit between wheat productivity and seasonal maximum temperature (**a**), and seasonal minimum temperature (**b**) in fourth agro-climatic zone

A significant relationship ($P = 0.05$) resulted from multiple linear regression analysis between seasonal weather elements and wheat productivity in the fourth agro-climatic zone, where R^2 value was 0.7297. Additionally, the goodness of fit analysis showed that the values of RMSE and *d-stat* between measured and predicted wheat productivity were 0.2126 ton/ha and 0.8276, respectively.

Although seasonal relative humidity exhibited the highest value of CV%, it did not have any significant contribution in wheat productivity in this zone. High values of R^2 between both seasonal maximum temperature, and seasonal minimum temperature with wheat productivity were found, exhibited a negative relationship (Fig. 5a, b). This result implied that contribution of seasonal temperature (both maximum and minimum) is more pronounced than the other two studied weather elements. It could be due to higher temperature values in this fourth agro-climatic zone, compared to the previous three zones.

Table 8 Descriptive statistics of seasonal weather elements in the fifth agro-climatic zone from 2010 to 2019

Statistic	SRAD	TMAX	TMIN	RH
Mean	18.94	27.40	11.76	32.50
Standard error	0.12	0.36	0.36	0.54
Standard deviation	0.39	1.15	1.14	1.70
Range	1.32	3.89	3.81	5.68
Minimum value	18.17	25.35	10.07	29.71
Maximum value	19.50	29.24	13.89	35.39
CV (%)	2.06	4.18	9.70	5.22

SRAD = solar radiation ($MJ/m^2/day$), TMAX = maximum temperature (°C), TMIN = minimum temperature (°C), and RH = relative humidity (%)

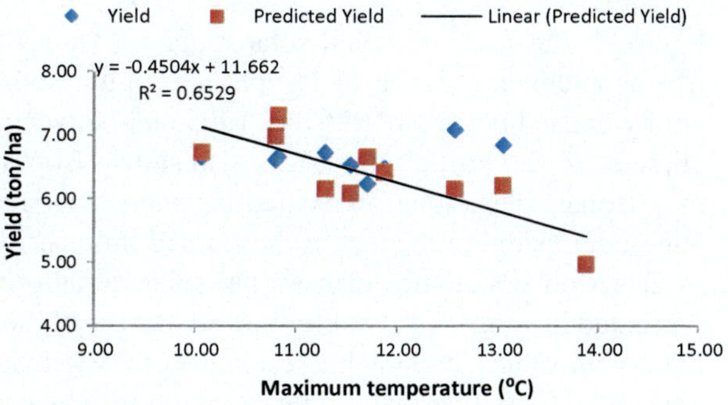

Fig. 6 Line fit between wheat productivity and seasonal minimum temperature in the fifth agro-climatic zone

The Fifth Agro-climatic Zone

Descriptive statistics analysis for the studied seasonal weather elements in the fifth agro-climatic zone indicated that the highest value of range between minimum and maximum values was found for seasonal relative humidity. The highest value of CV% was found for seasonal minimum temperature. Whereas, the lowest value of the range between minimum and maximum values and the lowest CV% were found for seasonal solar radiation (Table 8).

A significant relationship ($P < 0.05$) resulted from multiple linear regression analysis between seasonal weather elements and wheat productivity was found in the fifth agro-climatic zone, where R^2 value was 0.7015. The results of the goodness of fit analysis showed that RMSE value between measured and predicted wheat yield was 0.6459 ton/ha and *d-stat* value was 0.8109.

Even though higher value of CV% existed for the seasonal minimum temperature, it did not have any significant contribution in wheat productivity in this zone. High value of R^2 between seasonal maximum temperature and wheat productivity was found, which displayed a negative relationship (Fig. 6). This result implied that seasonal maximum temperature is the most important weather element that affect wheat productivity in the fifth agro-climatic zone.

9.2 *Assessment of Temporal-Variability of National Wheat Productivity and Climate*

In this assessment, the temporal-variability of national wheat productivity from year to year, as affected by the variability in each climate elements overall the agro-climatic zones was determined.

9.2.1 Seasonal Solar Radiation

Figure 7 illustrated seasonal solar radiation values between 1986 and 2019 in the five agro-climatic zones of Egypt. The figure showed that there were high variability existed between 1986 and 1993 and between 2008 and 2019 in all the agro-climatic zones. Furthermore, it is also shown from the figure that lower variability in seasonal solar radiation existed between 1984 and 2007 in all the agro-climatic zones, compared to the rest of the studied time period.

Descriptive statistics of seasonal solar radiation in the studied time period are presented in Table 9. The table showed that the highest range between minimum and maximum values, and the highest value of CV% existed in the second agro-climatic zone, which implied that higher variation in seasonal solar radiation existed in that zone, compared to the other zones. Whereas, the lowest value of the range was found in the third agro-climatic zone and the lowest value of CV% was found in the fourth agro-climatic zone.

Multiple linear regression analysis between seasonal solar radiation and national wheat productivity from 1986 to 2019 showed a highly significant value of R^2 equal to 0.5199, which means that 51.99% of the variation of national wheat productivity can be explained by seasonal solar radiation in that period.

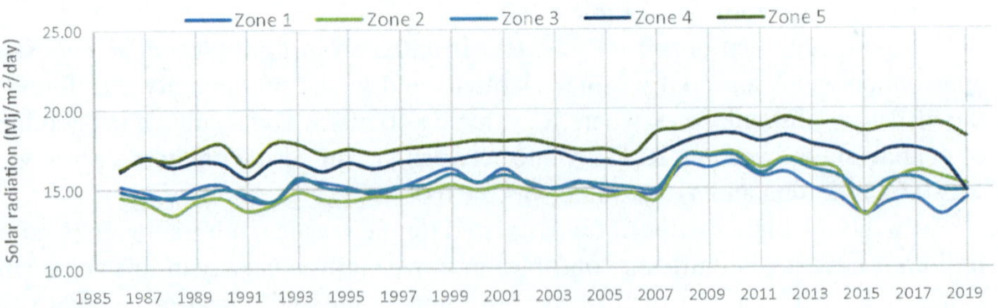

Fig. 7 Historical data of seasonal solar radiation from 1986 to 2019 in the agro-climatic zones of Egypt

Table 9 Descriptive statistics for seasonal solar radiation ($MJ/m^2/day$) data from 1986 to 2019 in the agro-climatic zones of Egypt

Statistic	Zone 1	Zone 2	Zone 3	Zone 4	Zone 5
Mean	15.00	15.06	15.35	16.92	17.99
Standard error	0.14	0.18	0.13	0.13	0.15
Standard deviation	0.81	1.05	0.75	0.75	0.88
Range	3.33	3.94	2.84	3.61	3.27
Minimum value	13.29	13.28	14.20	14.78	16.23
Maximum value	16.63	17.22	17.04	18.40	19.50
CV (%)	5.36	7.01	4.87	4.42	4.91

Furthermore, seasonal solar radiation was demonstrated as a statistically significant in the fifth agro-climatic zone only (Fig. 8), where it had the highest variation level, as shown by highest CV% value in Table 9. Furthermore, seasonal solar radiation was did not show any significant effect in the rest of the zones.

Morsy et al. (2017) stated that, in Egypt, solar radiation decreases gradually from south to north according to the apparent position of the sun and reaches its maximum value in the summer season, which explain the high contribution of seasonal solar radiation in national wheat productivity.

In an attempt to find more representative results to whether there was significant relationship between seasonal solar radiation and national wheat productivity in other zones, three time period were tested between 1968–1993, 1994–2007, and 2008–2019.

In the period between 1986 and 1993, multiple linear regression analysis showed that the national wheat productivity was insignificantly affected by seasonal solar radiation.

Whereas, in the period between 1994 and 2007, a highly significant relationship between national wheat productivity and seasonal solar radiation was obtained, where the value of R^2 ($P < 0.01$) was equal to 0.9112. This result demonstrated that 91.12% of the variation of national wheat productivity can be explained by seasonal solar

Fig. 8 Line fit between national wheat productivity and seasonal solar radiation in the fifth agro-climatic zone

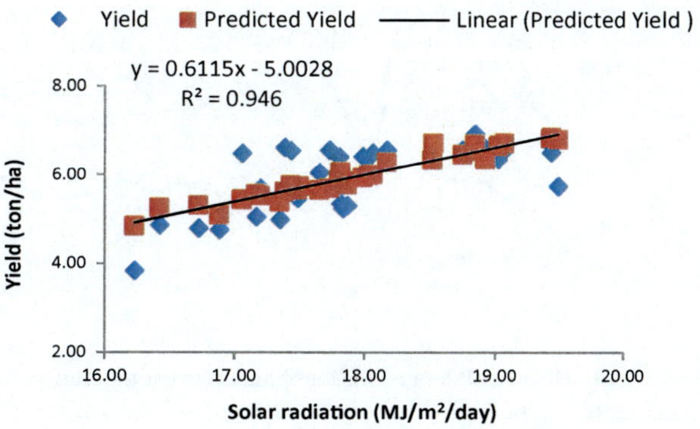

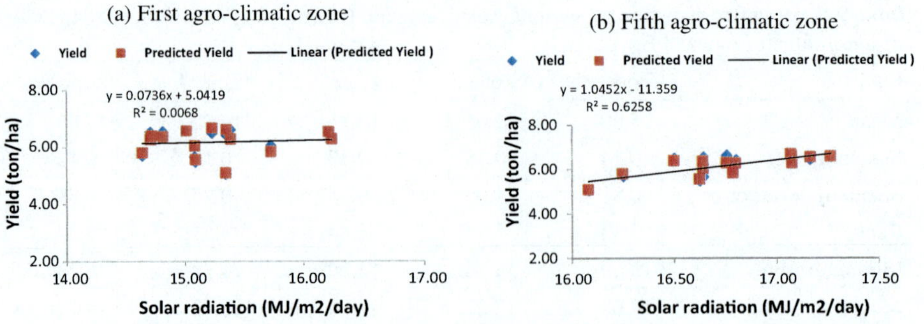

Fig. 9 Line fit between national wheat yield and seasonal solar radiation in first agro-climatic zone (**a**) and fourth agro-climatic zone (**b**) using data from 1994–2007

radiation in that time period. The goodness of fit analysis indicated that RMSE value between measured and predicted yield was 0.2060 ton/ha and *d-stat* value was 0.9513.

Moreover, a significant positive relationship between seasonal solar radiation and national wheat productivity in the above mentioned time period was found in the first and fourth agro-climatic zones ($P = 0.0126$ and $P = 0.0254$, respectively) (Fig. 9a, b).

Furthermore, in the period between 2008 and 2019, the analysis showed insignificant relationship between seasonal solar radiation and the national wheat productivity in the agro-climatic zones of Egypt.

9.2.2 Seasonal Maximum Temperature

With respect to seasonal maximum temperature, Fig. 10 showed high variability existed between the year of 1986 and 1996 and between the year of 2008 and 2019

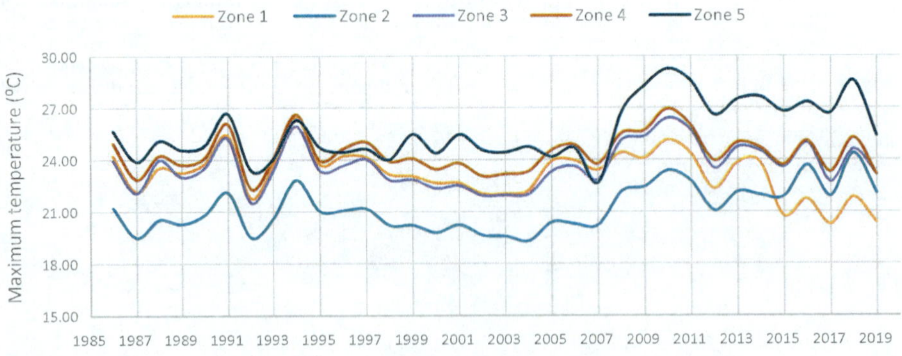

Fig. 10 Historical data of seasonal maximum temperature from 1986 to 2019 in the agro-climatic zones of Egypt

Table 10 Descriptive statistics for seasonal maximum temperature (ºC) data from 1986 to 2019 in the agro-climatic zones of Egypt

Statistic	Zone 1	Zone 2	Zone 3	Zone 4	Zone 5
Mean	21.16	23.17	23.63	24.36	25.63
Standard error	0.22	0.24	0.21	0.19	0.28
Standard deviation	1.31	1.39	1.25	1.09	1.64
Range	5.05	6.18	4.85	4.58	6.60
Minimum value	19.28	20.23	21.53	22.33	22.64
Maximum value	24.33	26.41	26.37	26.91	29.24
CV (%)	6.17	5.98	5.29	4.49	6.40

in all the agro-climatic zones of Egypt. Additionally, lower variably in seasonal maximum temperature existed between the year of 1997 and 2007.

Descriptive statistics analysis of seasonal maximum temperature in the period between 1986 and 2019 showed that the highest range minimum and maximum values and the highest value of CV% existed in the fifth agro-climatic zone, which implied high variation in the seasonal maximum temperature in the fifth agro-climatic zone. The lowest values of both statistics were found in fourth agro-climatic zone (Table 10).

Multiple linear regression analysis between seasonal maximum temperature and national wheat productivity in the period between 1986 and 2019 exhibited a significant value of R^2 ($P < 0.05$) equal to 0.5949.

In the second agro-climatic zone (Fig. 11), the value of R^2 between seasonal maximum temperature and national wheat productivity was 0.5114 and it was statistically significant ($P = 0.043$), with negative trend. Furthermore, seasonal maximum temperature values at the rest of the zones were statistically insignificant with R^2 values lower than 0.5000.

Moreover, regression analysis was done between seasonal maximum temperature and the national wheat productivity in three periods: between 1986 and 1996, between 1997 and 2007 and between 2008 and 2019.

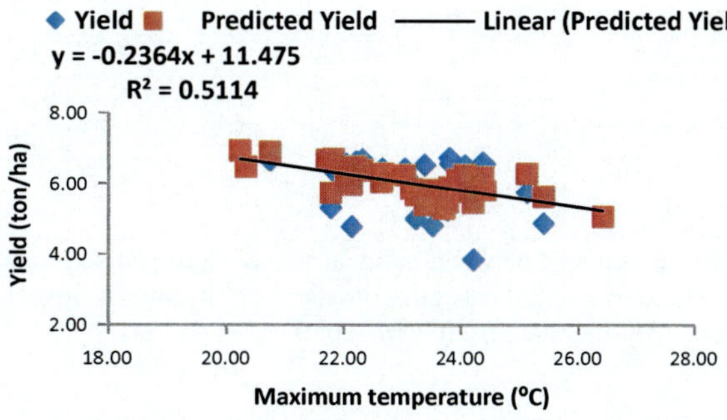

Fig. 11 Line fit between national wheat productivity and seasonal maximum temperature in the second agro-climatic zone

The analysis showed that, in the period between 1986 and 1996, as well as in the period between 1979 and 2007, there were insignificant relationship between seasonal maximum temperature and the national wheat productivity.

However, in the period between 2008 and 2019, the relationship between seasonal maximum temperature and the national wheat productivity was found significant (*P* < *0.05*), where R^2 value was 0.7132. This result demonstrated that 71.32% of the variation of national wheat productivity can be explained by seasonal maximum temperature in this time period. The goodness of fit analysis between seasonal maximum temperature and national wheat productivity showed low value of RMSE, namely 0.2027 ton/ha and high value of *d-stat,* namely 0.8130.

Furthermore, a significant relation between seasonal maximum temperature and national wheat productivity was found for all zones (Fig. 12), except in the second agro-climatic zone. Figure 12 also showed that high R^2 values between seasonal maximum temperature and national wheat productivity were found, which implied a strong relationship between them in the period between 2008 and 2019 in all agro-climatic zones, except in the second agro-climatic zone. Morsy et al. (2017) indicated that maximum temperature in Egypt increases gradually southward in all seasons following the apparent position of the sun.

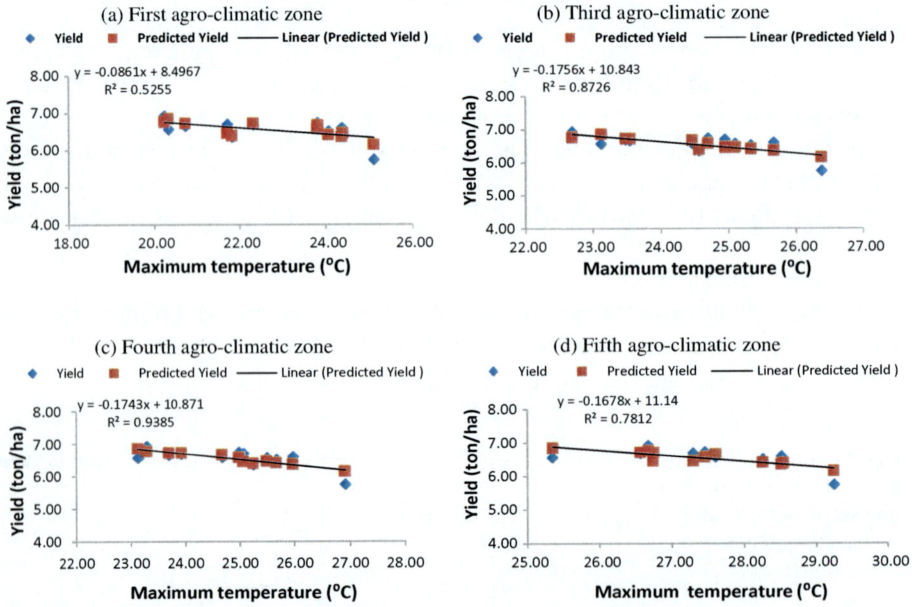

Fig. 12 Line fit between national wheat yield and seasonal maximum temperature in first agro-climatic zone (**a**), third agro-climatic zone (**b**), fourth agro-climatic zone (**c**) and fifth agro-climatic zone (**d**) using data from 2008–2019

9.2.3 Seasonal Minimum Temperature

Similar trend to the one found for seasonal maximum temperature was found for seasonal minimum temperature (Fig. 13), where high variability existed between the year of 1986 and 1996 and between the year of 2008 and 2019 in all the agro-climatic zones of Egypt and lower variability existed between the year of 1997 and 2007.

The highest range between minimum and maximum values and CV% value of seasonal minimum temperature was found in the fifth agro-climatic zone. The lowest value of the range was found in the second agro-climatic zone and the lowest value of CV% was found in the first agro-climatic zone (Table 11).

Multiple linear regression analysis between seasonal minimum temperature and national wheat productivity in the period between 1968 and 2019 showed a highly significant value of R^2 ($P < 0.01$) equal to 0.8128.

In the fourth agro-climatic zone, the value of R^2 between seasonal minimum temperature and national wheat yield was 0.4161 and it was statistically significant ($P = 0.039$) (Fig. 14a). Similarly, in fifth agro-climatic zone, the value of R^2 between seasonal minimum temperature and national wheat yield was 0.2837 and it was statistically significant ($P = 0.048$) (Fig. 14b). Furthermore, the values of minimum

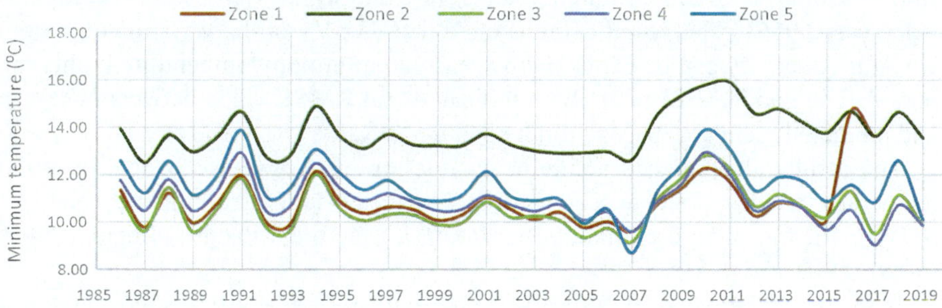

Fig. 13 Historical data of seasonal minimum temperature from 1986 to 2019 in the agro-climatic zones of Egypt

Table 11 Descriptive statistics for seasonal minimum temperature (°C) data from 1986 to 2019 in in the agro-climatic zones of Egypt

Statistic	Zone 1	Zone 2	Zone 3	Zone 4	Zone 5
Mean	13.77	10.75	10.51	10.85	11.53
Standard error	0.15	0.13	0.15	0.15	0.18
Standard deviation	0.90	0.77	0.90	0.87	1.07
Range	3.38	2.80	3.66	3.93	5.21
Minimum value	12.50	9.56	9.14	9.00	8.67
Maximum value	15.88	12.36	12.80	12.93	13.89
Coefficient of variation	6.55	7.21	8.57	8.04	9.29

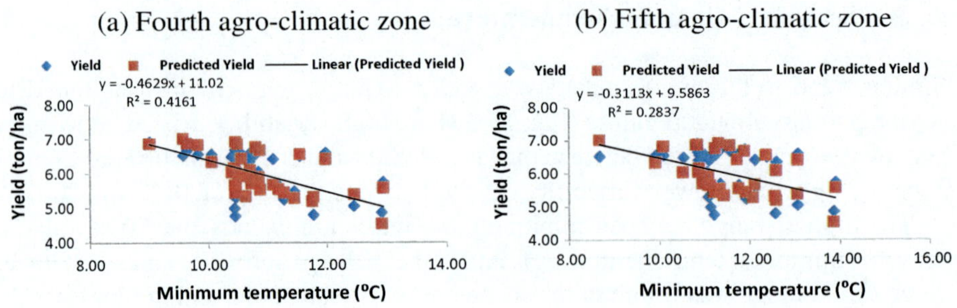

Fig. 14 Line fit between national wheat yield and seasonal minimum temperature in the fourth agro-climatic zone (**a**) and fifth agro-climatic zone (**b**)

temperature at the rest of the zones were statistically insignificant with R^2 values lower than 0.5000.

Morsy et al. (2017) reported that most of Middle and Upper Egypt, where fourth and fifth agro-climatic zones are located, have the lowest values of minimum temperature.

In the period between 1986 and 1996, the analysis showed a significant relationship ($P < 0.05$) between minimum temperature and wheat yield, where multiple R^2 value was 0.9561. This result demonstrated that 95.61% of the variation of national wheat productivity can be explained by seasonal minimum temperature in this time period. The goodness of fit analysis indicated that RMSE value between measured and predicted yield was 0.1442 ton/ha and *d-stat* value was 0.9771.

The results also showed a significant relationship between seasonal minimum temperature and national wheat productivity in the first agro-climatic zone ($P < 0.05$), in the fourth agro-climatic zone ($P < 0.01$) and the fifth agro-climatic zone ($P < 0.01$) (Fig. 15).

In the period between 1979 and 2007, the analysis indicated a significant relationship ($P < 0.05$) between seasonal minimum temperature and national wheat productivity, where R^2 value was 0.9445. RMSE value between measured and predicted wheat yield was 0.0934 ton/ha and *d-stat* value was 0.9707 as shown by the goodness of fit analysis.

A significant relationship between seasonal minimum temperature and national wheat productivity was found in the third agro-climatic zone ($P < 0.05$), in the fourth agro-climatic zone ($P < 0.05$) and in the fifth agro-climatic zone ($P < 0.05$) (Fig. 16).

Additionally, in the period between 2008 and 2019, the relationship between seasonal minimum temperature and the national wheat productivity was found highly significant ($P < 0.01$), where multiple R^2 value was 0.9612. Goodness of fit analysis indicated that RMSE value between measured and predicted yield was 0.0797 ton/ha and *d-stat* value was 0.9799.

Furthermore, a significant relationship between seasonal minimum temperature and national wheat productivity was found in the second agro-climatic zone ($P < 0.01$) and in the fourth agro-climatic zone ($P < 0.05$) (Fig. 17).

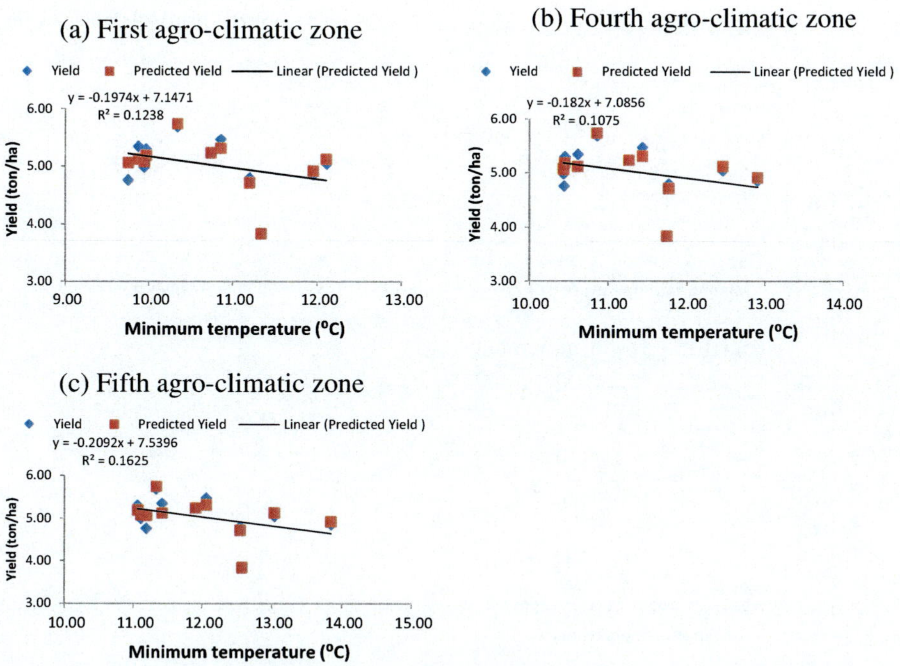

Fig. 15 Line fit between national wheat yield and seasonal minimum temperature in the first agro-climatic zone (**a**), in the fourth agro-climatic zone (**b**), and in the fifth agro-climatic zone (**c**) using data from 1986–1996

9.2.4 Seasonal Relative Humidity

Regarding seasonal relative humidity, Fig. 18 showed low variability throughout the studied data set, which could imply low impact on national wheat productivity.

The highest range between minimum and maximum values and CV% of seasonal relative humidity existed in the fourth agro-climatic zone. The lowest value of the range was found in the fifth agro-climatic zone and the lowest value of CV% was found in the first agro-climatic zone (Table 12).

Multiple linear regression analysis between seasonal relative humidity and national wheat productivity showed a highly significant value of R^2 ($P < 0.01$) equal to 0.7516.

In Zone 5, the value of R^2 was 0.4871 and it was highly statistically significant ($P = 0.0002$), with decreasing trend (Fig. 19). Furthermore, the values of seasonal relative humidity at the rest of the zones were statistically insignificant with R^2 values lower than 0.5000.

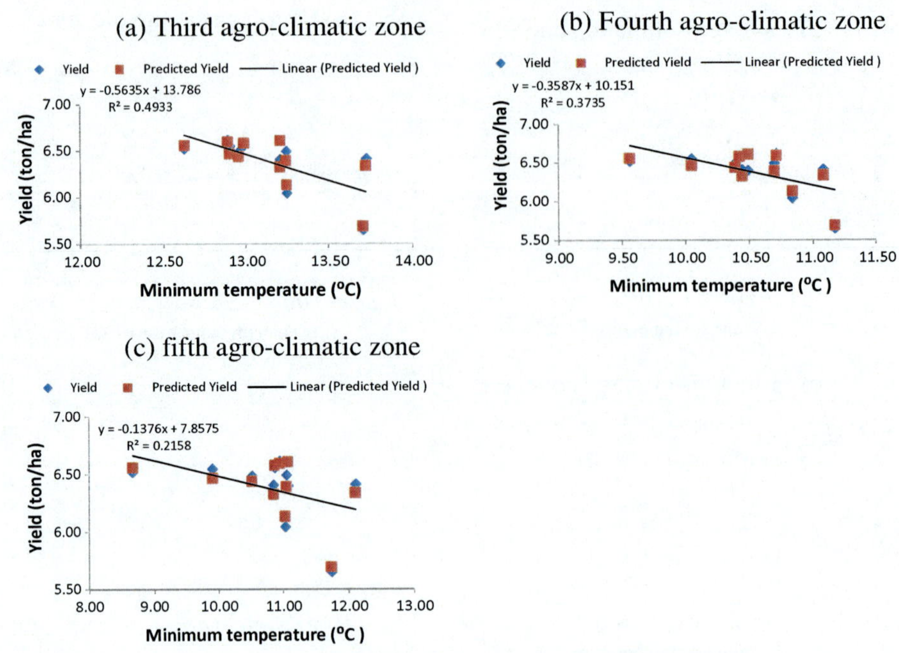

Fig. 16 Line fit between national wheat yield and seasonal minimum temperature in the third agro-climatic zone (**a**), in the fourth agro-climatic zone (**b**), and in the fifth agro-climatic zone (**c**) using data from 1979–2007

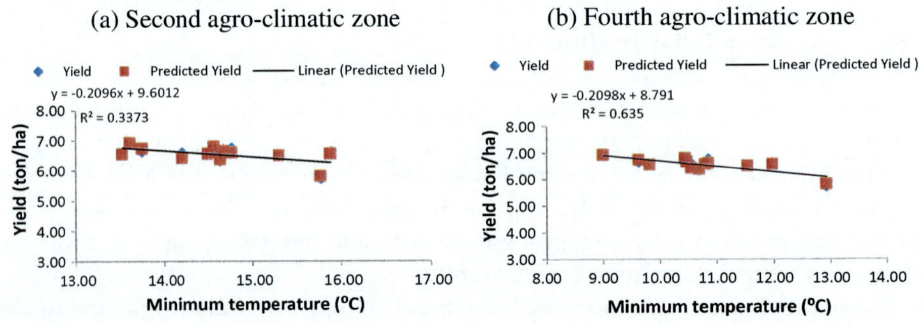

Fig. 17 Line fit between national wheat yield and seasonal minimum temperature in second agro-climatic zone (**a**), and in the fourth agro-climatic zone (**b**) using data from 2008–2019

10 Conclusions

Weather patterns causing huge uncertainty in crops production and consequently causing food insecurity. Because wheat is the major crop in the food basket of the Egyptian population and there is a large gap between its production and consumption, it is important to test whether the Egyptian wheat cultivars are resilient to climate

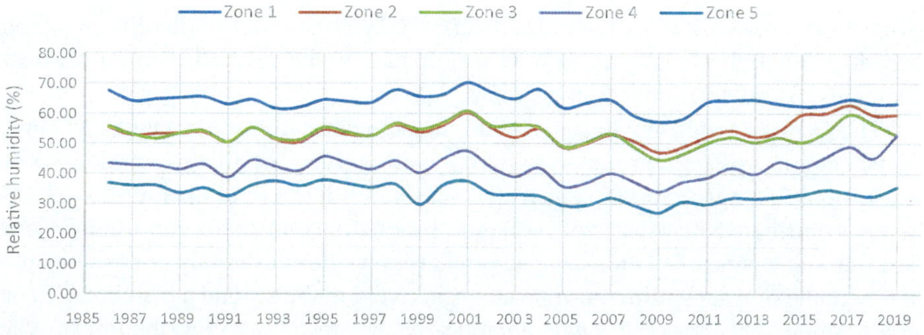

Fig. 18 Historical data of seasonal relative humidity from 1986 to 2019 in the agro-climatic zones of Egypt

Table 12 Descriptive statistics for seasonal relative humidity data from 1986 to 2019 in the agro-climatic zones of Egypt

Statistic	Zone 1	Zone 2	Zone 3	Zone 4	Zone 5
Mean	64.06	54.08	53.10	42.03	33.54
Standard error	0.45	0.63	0.59	0.65	0.49
Standard deviation	2.65	3.68	3.46	3.81	2.84
Sample variance	7.02	13.56	12.00	14.55	8.06
Range	13.01	15.91	16.37	18.84	10.78
Minimum value	57.26	46.96	44.51	33.96	27.09
Maximum value	70.28	62.88	60.88	52.80	37.87
CV (%)	4.14	6.81	6.52	9.08	8.47

Fig. 19 Line fit between national wheat productivity and seasonal relative humidity in the fifth agro-climatic zone

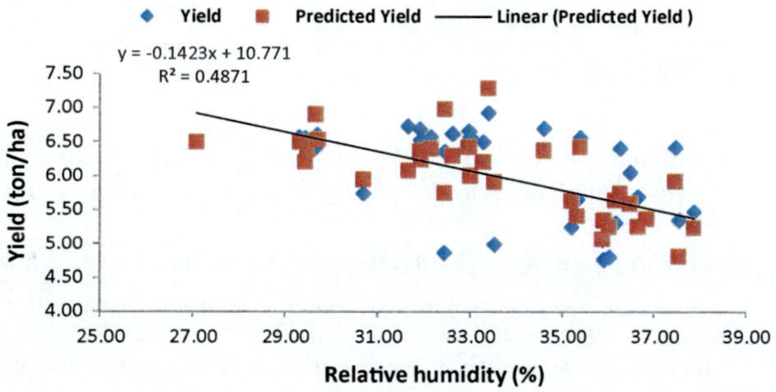

variability and possess yield stability traits under the experience adverse weather conditions prevailed lately.

Multiple regression analysis was used to find out which weather element have contributed more in wheat productivity and how strong is the contribution. The analysis determined the spatial-variability of wheat zone-specific productivity from one agro-climatic zone to another, as a result of the studied weather elements in each zone. Furthermore, the analysis also determined the temporal-variability of national wheat productivity from year to year, as affected by the variability in each weather elements overall the agro-climatic zones.

The results showed that seasonal maximum temperature was the major contributor in wheat zone-specific productivity in a negative trend, in all agro-climatic zones, except the first agro-climatic zone. The seasonal minimum temperature also negatively contributed in wheat zone-specific productivity in the second, third, fourth agro-climatic zones. Whereas, the seasonal relative humidity only contributed positively in the first agro-climatic zone and negatively in the second agro-climatic zone. Furthermore, the seasonal solar radiation did not have any contribution in wheat zone-specific productivity.

Furthermore, the analysis of the temporal-variability of national wheat productivity from year to year, as affected by the variability in each weather elements overall the agro-climatic zones revealed that there was higher variability existed in the studied climate elements in the past 10–12 years, compared to the rest of the data set. Furthermore, the analysis also showed that the seasonal solar radiation had a positive significant contribution in national wheat productivity in the first and fourth agro-climatic zones. The seasonal maximum and minimum temperatures had a negative significant contribution in all the agro-climatic zones. Whereas, the seasonal relative humidity had a negative significant contribution in the fifth agro-climatic zone only.

Thus, these results implied that the contribution of seasonal weather elements on zonal level, is different than its contribution of on the national level. Furthermore, these results ascertained that the Egyptian wheat cultivars are resilient to climate variability and possess yield stability traits because it have been exposed to unfavorable weather condition caused variability in its production, with insignificant reduction.

References

Abdelmageed K, Chang X, Wang D, Wang Y, Yang Y, Zhao G, Tao Z (2019) Evolution of varieties and development of production technology in Egyptian wheat: a review. J Integr Agric 18(3):483–495

Abou GA, Tabl KM (2014) Heat tolerance in some Egyptian wheat cultivars as measured by membrane thermal stability and proline content A. Middle East J Agric Res 3(2):186–193

Ahmed M, Fayyaz-ul-Hassan A, Aslam MA, Akram MN, Akmal M (2011) Regression model for the study of sole and cumulative effect of temperature and solar radiation on wheat yield. African J Biotechnol 10(45):9114–9121

Almeselmani M, Deshmukh PS, Sairam RK (2009) High temperature stress tolerance in wheat genotypes: role of antioxidant defense enzymes. Acta Agron Hungar 57:1–14

Ambavaram MMR, Basu S, Krishnan A, Ramegowda V, Batlang U, Rahman L, Baisakh N, Pereira A (2014) Coordinated regulation of photosynthesis in rice increases yield and tolerance to environmental stress. Nature 36:93–93

Blum A (2011) Plant breeding for water-limited environments, 1st edn. Springer, New York, NY, USA

Chen YH, Yu SL, Yu ZW (2003) Relationship between amount or distribution of PAR interception and grain output of wheat communities. Acta Agron Sin 29:730–734

Craufurd PQ, Wheeler TR (2009) Climate change and the flowering time of annual crops. J Exp Bot 60:2529–2539

Deng NY, Ling XX, Sun Y, Zhang CD, Fahad S, Peng SB, Cui KH, Nie LX, Huang JL (2015) Influence of temperature and solar radiation on grain yield and quality in irrigated rice system. Eur J Agron 64:37–46

Draper NR, Smith H (1987) Applied regression analysis. Wiley, New York, pp 397–402

Draz I, Esmail S, Abou-Zeid M, Hafez Y (2018) Changeability in stripe rust infection and grain yield of wheat associated to climatic conditions. Env Biodiv Soil Secur 2:43–153

Draz IS, Abou-Elseoud MS, Kamara AM, Alaa-Eldein OA, El-Bebany AF (2015) Screening of wheat genotypes for leaf rust resistance along with grain yield. Ann Agric Sci 60(1):29–39

Farooq M, Wahid A, Kobayashi N, Fujita D, Basra SMA (2009) Plant drought stress: effects, mechanisms and management. Agron Sustain Dev 29:185–212

Gomez KA, Gomez AA (1984) Statistical procedures for agricultural research, 2nd edn. Willey, Toronto, ON, Canada

Gouache D, Le Bris X, Bogard M, Deudon O, Page C, Gate P (2012) Evaluating agronomic adaptation options to increasing heat stress under climate change during wheat grain filling in France. Eur J Agron 39:62–70

Grant RF, Kimball BA, Conley MM, White JW, Wall GW, Ottman MJ (2011) Controlled warming effects on wheat growth and yield: field measurements and modeling. Agron J 103(6):1742–1754. https://doi.org/10.2134/agronj2011.0158

Han H, Li H, Ning T, Zhang X, Shan Y, Bai M (2008) Radiation use efficiency and yield of winter wheat under deficit irrigation in North China. Plant Soil Environ 54(7):313–319

Hoggs WH, Houman CE, Mallik AK, Zadoks JC (1969) Meteorological factors affecting the epidemiology of wheat rusts. World Meteorological Organization, Geneva, Switzerland

Hossain A, Sarker MAZ, Saifuzzaman M, Teixeira da Silva JA, Lozovskaya MV, Akhter MM (2013) Evaluation of growth, yield, relative performance and heat susceptibility of eight wheat (Triticum aestivum L.) genotypes grown under heat stress. Int J Plant Prod 7:615–636

Howarth CJ (2005) Genetic improvements of tolerance to high temperature. In: Ashraf M, Harris PJC (eds) Abiotic stresses: plant resistance through breeding and molecular approaches. Howarth Press Inc., New York

Jamieson PD, Porter JR, Goudriaan J, Ritchie JT, Keulen H, Stol W (1998) A comparison of the models AFRCWHEAT2, CERES-Wheat, Sirius, SUCROS2 and SWHEAT with measurements from wheat grown under drought. Field Crops Res 55:23–44

Kaushal N, Bhandari K, Siddique KHM, Nayyar H (2016) Food crops face rising temperatures: an overview of responses, adaptive mechanisms, and approaches to improve heat tolerance. Cogent Food Agric 2(1):1134380. https://doi.org/10.1080/23311932.2015.1134380

Khalil AA, Hassanein MK (2016) Extreme weather events and negative impacts on Egyptian agriculture. Int J Adv Res 4(12):1843–1851

Kosova K, Vitamvas P, Prasil IT, Renaut J (2011) Plant proteome changes under abiotic stress-contribution of proteomics studies to understanding plant stress response. J Proteome 74:1301–1322. https://doi.org/10.1016/j.jprot.2011.02.006

Lecoeur J, Ney B (2003) Change with time in potential radiation use efficiency in field pea. Eur J Agron 19:91–105

Lobell DB, Cassman KG, Field CB (2009) Crop yield gaps: their importance, magnitudes, and causes. Ann Rev Environ Resour 34:179–204

Lukac M, Gooding MJ, Griffiths S, Jones HE (2011) Asynchronous flowering and within-plant flowering diversity in wheat and the implications for crop resilience to heat. Ann Bot 109:843–850. https://doi.org/10.1093/aob/mcr308

Lyon B, Broders K (2017) Impact of climate change and race evolution on the epidemiology and ecology of stripe rust in Central and Eastern USA and Canada. Can J Plant Path 39(4):385–392

Ministry of Agriculture and Land Reclamation (2019) Agricultural statistical bulletin. Part two: Summer crops. Economic Affairs Sector, Egypt
Morsy M, Sayad T, Ouda S (2017) Present and future water requirements for crops. In: Future of food gaps in Egypt: obstacles and opportunities. Springer Publishing House
Nahar K, Gretzmacher R (2002) Effect of water stress on nutrient uptake, yield and quality of tomato (Lycopersicon esculentum Mill.) under subtropical conditions. Die Bodenkultur 53(1):45–51
Oshino T, Miura S, Kikuchi S, Hamada K, Yano K, Watanabe M, Higashitani A (2011) Auxin depletion in barley plants under high temperature conditions represses DNA proliferation in organelles and nuclei via transcriptional alterations. Plant Cell Environ 34:284–290. https://doi.org/10.1111/j.1365-3040.2010.02242.x
Ouda S, Noreldin T (2017) Evapotranspiration data to determine agro-climatic zones in Egypt. J Water Land Dev 32(I-III):79–86
Prasad PVV, Djanaguiraman M (2014) Response of floret fertility and individual grain weight of wheat to high temperature stress: sensitive stages and thresholds for temperature and duration. Funct Plant Biol 41:1261–1269. https://doi.org/10.1071/FP14061
Ren XL, He HL, Zhang L, Zhou L, Yu GR, Fan JW (2013) Spatiotemporal variability analysis of diffuse radiation in China during 1981–2010. Ann Geophys 31:277–289. https://doi.org/10.5194/angeo-31-277-2013
Royo C, Maccaferri M, Álvaro F, Moragues M, Sangquineti MC, Tuberosa R, Maalouf F, García del Moral LF, Demontis A, Rhouma S, Nachit M, Nserallah N, Villegas D (2010) Understanding the relationships between genetic and phenotypic structures of a collection of elite durum wheat accessions. Field Crops Res 119:91–105
Teixeira EI, Fischer G, Velthuizen HV, Walter C, Ewert F (2013) Global hot spots of heat stress on agricultural crops due to climate change. Agric for Meteorol 170:206–215. https://doi.org/10.1016/j.agrformet.2011.09.002
Villegas D, Alfaro C, Ammar K, Cátedra MM, Crossa J, García del Moral LF, Royo C (2015) Daylength, temperature and solar radiation effects on the phenology and yield formation of spring durum wheat. J Agron Crop Sci 202(3):203–216. https://doi.org/10.1111/jac.12146
Wang X, Cai J, Jiang D, Liu F, Dai T, Cao W (2011) Pre-anthesis high temperature acclimation alleviates damage to the flag leaf caused by post-anthesis heat stress in wheat. J Plant Physiol 168:585–593. https://doi.org/10.1016/j.jplph.2010.09.016
Wassmann R, Jagadish SVK, Heuer S, Ismail A, Redona E, Serraj R, Singh RK, Howell G, Pathak H, Sumfleth K (2009) Climate change affecting rice production. Adv Agron 101:59–122
Wiese MV (1987) Compendium of wheat diseases, 2nd edn. American Phytopathological Society, St. Paul, MN
Willmott CJ (1981) On the validation of models. Phys Geogr 2:184–194
Wolfenden R (2003) Thermodynamic and extra thermodynamic requirements of enzyme catalysis. Biophys Chem 105:559–572
World Meteorological Organization (2012) Standardized precipitation index user guide. World Meteorological Organization, Geneva
Zhang XY, Wang SF, Sun HY, Chen SY, Shao LW, Liu XW (2013) Contribution of cultivar, fertilizer and weather to yield variation of winter wheat over three decades: a case study in the North China Plain. Eur J Agron 50:52–59
Zhao C, Liu B, Piao S et al (2017) Temperature increase reduces global yields of major crops in four independent estimates. PNAS 114:9326–9331
Zohry A, Ouda S (2018) Cropping pattern to increase food security. In: Cropping pattern to overcome abiotic stresses: water, salinity and climate. Springer Publishing House
Zohry A, Ouda S (2020) Wheat insufficiency and deficit irrigation. In: Deficit irrigation: a remedy for water scarcity. Springer Publishing House

Part III
Reduction of Greenhouse Gases

Chapter 8
Practices Contribute in Reducing the Emission of Greenhouse Gases

Samiha Ouda and Abd El-Hafeez Zohry

Abstract Both agriculture and food systems have influenced climate change and vice versa. In this chapter, some of the implication of climate change on agriculture, internationally and locally were reviewed. The sources of emission of greenhouse gases in agriculture, namely emission from soil, from rice fields and burning of crops residues were discussed. The climate-smart practices that could be implemented to reduce the emission of greenhouse gases from these three sources, such as carbon sequestration in the soil, reduction of the emission from rice fields and biogas production from crops residues will be also reviewed.

Keywords Climate change · Greenhouse gasses emission from the soil · From greenhouse gasses emission rice fields · Greenhouse gasses emission from burning crop residues

1 Introduction

Climate change is referred to as "change in the state of the climate that can be identified (e.g., by using statistical tests) by changes in the mean and/or the variability of its properties and persisting for an extended period, typically decades or longer" (IPCC 2014). The Inter-Governmental Panel on Climate Change (IPCC) in its Fifth Assessment Report has warned that global climate has been changing and it would continue to happen in the near future, where the global mean surface temperatures are projected to increase by 1.4–5.8 °C by the end of this century relative to 1990 (IPCC 2014). Global warming phenomenon appears as a widespread rising of surface air temperatures, alteration of precipitation patterns and global hydrologic cycle that

S. Ouda (✉)
Water Requirements and Field Irrigation Research Department, Soils, Water and Environment Research Institute; Agricultural Research Center, Giza, Egypt

A. E.-H. Zohry
Crops Intesefication Research Department, Field Crops Research Institute; Agricultural Research Center, Giza, Egypt

S. Ouda and A. E.-H. Zohry, *Climate-Smart Agriculture*,
https://doi.org/10.1007/978-3-030-93111-7_8

increased the frequency of severe weather events such as drought spells and floods leading to low water availability (Mo et al. 2013).

There is substantial evidence to show that climate change is already affecting agricultural production systems worldwide. Global climate models predicted 1–6 °C increase in the mean ambient temperature in the end of twenty-first century (De Costa 2011). Such increase may have a significant negative influence on agricultural productivity (Bates et al. 2008), in addition to increasing drought, salinity, waterlogging, and mineral toxicity stresses (Teixeira et al. 2013). These changes will have a dramatic effect on agricultural systems especially in temperate countries, where the water shortage and high temperatures are the main limiting factors for the crop production (Rinaldi 2009). Therefore, adaptation measures should be applied to reduce the harmful effects of climate change. Adaptation is defined as the "decision-making process and the set of actions undertaken to maintain the capacity to deal with future changes or perturbations to socio-ecological system without undergoing significant changes in function, structural identity, while maintaining the option to develop" (Nelson et al. 2007).

Human activities, such as alterations in land usage patterns (deforestation, tillage of the soil, application of fertilizers, burning of biomass, extrication of crop leftovers) and its consequential greenhouse gases emission is the main reasons for the negative climate change phenomena (IPCC 2018). Burning fossil fuels added substantive amounts of carbon dioxide (CO_2), led to continuance in the generation of greenhouse gases. The greenhouse gases are CO_2, methane (NH_4), nitrous oxide (N_2O), chlorofluorocarbons, etc. The high release of these gases through the time has thickened the greenhouse gases layer and trap it within the earth's atmosphere, disturbing the earth's thermostat (Singh and Khanna 2021). Some studies reported that about 45% of human activities release CO_2 remains in the atmosphere. It has been found that around 77% of all emissions are CO_2 emissions, 14% are NH_4, 8% are N_2O and 1% is chlorofluorocarbon and others (World Resource Institute 2000). IPCC (2018) indicated that agriculture, land use and forestry contribute in emitting around 14% and 6% of the greenhouse gases, respectively. The increased greenhouse gas emissions have also been responsible for the loss of biodiversity (Knox et al. 2016).

Agriculture is deeply intertwined with food systems. Both agriculture and food systems have impacted climate change and vice versa. A food system includes all activities right from the input supply, manufacturing and production, harvesting, processing, packaging, marketing/distribution and consumption, i.e., from farm to fork (Singh and Khanna 2021). A food system is linked with the ecosystems as land and water are utilized at every stage to produce food and it produces waste too, which is disposed back into the land and water bodies and emits greenhouse gases, also polluting water bodies, and resulting in soil degradation (Mbow et al. 2019). Food systems contribute approximately 19–29% to the world's greenhouse gases out of which 80–86% is contributed by agricultural activities (Vermeulen et al. 2012).

Several smart practices can be implemented to reduce the emission of the greenhouse gases, mainly CO_2 and methane. Regarding CO_2, practices that maintain soil cover year around can sequestrate CO_2 in the soil and prevent its emission. Some of these practices are implementing intercropping systems, cultivation of three crops per year, and implementing crop rotations. Similarly, to reduce methane emission, matching application of nitrogen fertilizer to crops with plant needs, and replace mineral fertilizer with organic fertilizer. Furthermore, use farms wastes in producing biogas to reduce dependence on fossil fuel.

In this chapter we will review some of the implication of climate change on agriculture. The practices that could be implemented to reduce the emission of greenhouse gases to attain the third pillar of smart-climate agriculture will be also reviewed.

2 Implication of Climate Change on Agriculture

2.1 Effects on Water Resources

Climate change and its impact on water resources are a further challenge in the semi-arid regions as they are fragile ecosystems. As stated by IPCC (2007), climate change will have a negative impacts on water resources and freshwater availability, as well as the hydrological cycle in most of the regions of the world (Wu et al. 2014). Rainfall variability is expected to increase, as well as the incidence of extreme weather conditions, such as floods, droughts and changing rainfall patterns (IPCC 2014). All these events will lead to a loss of crops and food shortage (Liwenga et al. 2007). Variations of groundwater recharge, as a result of shifts in precipitation and temperature could cause water table fluctuations (Abbaspour et al. 2009). Ali et al. (2012) reported that changes in groundwater recharge can have an influence on evapotranspiration, groundwater flow direction, groundwater level and surface water–ground water interaction. Xing-Guo et al. (2017) projected that, by the 2050s, the increased crop water demand and intensified evapotranspiration resulting from global warming will reduce water resources surplus by about 4–24% and increase significantly the irrigation water demand in crop growth periods.

2.2 Effects on Crops Production

Eyshi Rezaei et al. (2018) reported that acceleration in crop development and phenological stages are an important bio-indicators of climate change. Other disadvantages are expected as a result of exposure to climate change, namely an increase in maintenance respiration and reduction in stomatal conductance causing variation of yield and water use efficiency (Mo et al. 2013). On the other hand, alteration of precipitation will affect water availability for crops, so that crop yield will be severely affected or even crop failure could occur (Asseng et al. 2013). Emam et al. (2015) projected that, as a result of decrease in the mean annual precipitation, soil water content will be decreased and that will resulted in the reduction of crop yield in rain-fed areas. Thus, global climate change adversely impact crop production and imposes a wide range of constraints on agricultural systems especially in water-limited environments (Paymard et al. 2018). It was estimated that up to one fifth of the global population could suffer severe shortages of fresh water (Schiermeier 2014) or quality water (Girones et al. 2010) in the future. Furthermore, high losses in the yield of several field crops are reported depending on the location, future climate change and projected yield (Porter et al. 2014), which ascertain that climate change will be a challenge for global food security.

Furthermore, the exposure of roots to temperatures higher than the optimal causes a decrease in the primary root length, number of lateral roots and their angle of emergence, in addition to the initiation of second and third order lateral roots that are characterized by a larger diameter (Calleja-Cabrera et al. 2020). The negative effect of increasing temperatures usually reduces the surface between root and soil, therefore decreasing nutrient and water uptake (Nagel et al. 2009). In wheat, seminal and crown roots retarded their emergence and elongation when its seedlings are grown under higher temperature than optimal (Huang et al. 1991). Furthermore, the increase in temperature causes a decrease in the length and number of central late metaxylem in the root tip in wheat. Morales et al. (2003) indicated that this change could be considered as an adaptation to limit injury in the root by the changes in water viscosity and root hydraulic conductance produced by heat stress. Whereas, in maize adult plants, the increase in temperature slows down lateral root growth to promote the development of long axile roots to reach the water of the deeper soil layers (Hund et al. 2008). Joshi et al. (2016) reported that alteration of root growth in these crops seems to be caused by a decrease in the cell division rate.

Regional studies of climate change impacts on staple crops in the United States, such as wheat (Urban et al. 2015) and maize (Challinor et al. 2010) were done. These studies projected an increase in mean temperature will lead to rising yield variability and incidences of crop failure. With continued warming under business-as-usual greenhouse gas emissions, global crop yields are expected to decline significantly. Parry et al. (2004) reported a reduction of rain fed and irrigated wheat yield by 10–40 and 5–20%, respectively. Wheat grown in warmer regions is likely to experience greater yield losses than that grown in cooler regions, though there is also general agreement that high latitude spring wheat production will benefit from a warmer climate through an extension of the growing period (Sommer et al. 2013). Liu and Yang (2010) stated that 2–5 °C increase in temperature in China could have more effects on wheat yield than the rise in CO_2 concentration, although it is mainly a negative impact. Whereas, Zhao et al. (2017a, b) reported that for every degree increase in global mean temperature, wheat yield are projected to decrease, on average, by 6%.

With respect to maize, it is more sensitive to high temperature stress, compared to wheat in spite of being a warm season crop (Tesfaye et al. 2015). Higher temperatures will have negative impacts on maize yields with multiplication of some weeds and pests in the same time (Shiferaw et al. 2011). It was reported that at a high temperature of 35 °C, maize yield reduces by 9% with rainfall reduction (Adhikari et al. 2015). Msowoya et al. (2016) stated that rain fed maize production in Malawi, may decrease up to 14% by mid-century due to climate change, rising to as much as 33% loss by the end of the century. Zhao et al. (2017a, b) projected reduction in maize yield, on average, by 7% for every degree increase in global mean temperature. Thus, even though new maize hybrids are developed by plant breeders adopted to biophysical environmental stresses, it might not be able to adopt to climate change stresses and its productivity still diminished (Banziger and Diallo 2004).

2.3 *Effects on Agricultural Soils*

Soil is the medium for cultivating plants, in which nutrients and water are provided for their growth. Climate change will impact soil through intermittent rainfall patterns and frequent drought, which negatively affect soil supplying capacity of water and nutrients to the growing plants (Faurès et al. 2013). The risk of soil erosion by heavy rainfall is predominant of climate change. A rise in the temperature can increased the rate of soil organic matter mineralization (Lal 2015). Changes in temperature and precipitation will change the diversity and functionality of fungal communities (Chanda et al. 2020). Warming may stimulate carbon losses from the soil by increasing rates of decomposition by the saprophytes fungi. The saprophytic activity can be influenced directly through variations in temperature, precipitation or by indirect effect as variations in soil moisture, which mobilizing the large carbon stocks in soil, and increasing soil respiration and decomposition (Allison and Treseder 2011). The variation of precipitation and temperature can lead to land degradation, especially in fragile arid and semi-arid ecosystems (Meadows and Hoffman 2003).

2.4 *Effects on Water Consumption by Crops*

With the projection of increase in the average global temperature, which may increase by 1.4–5.8 °C, another projection was stated that there would be a substantial reduction in freshwater resources as a result of rainfall variability (Tadross and Wolski 2010). Increased temperatures lead to increased evapotranspiration, which affects water availability for crops (Holmén 2003). In some parts of the world, the projected climate change is likely to increase water stress and demand for increased irrigation, hence leading to a drop in crop yields (Ravindranath et al. 2010). Rotich and Mulungu (2017) projected water requirements of maize planted in two sites in Tanzania and they reported an increase by 4% in the 2020s and 7% in the 2050s at the Moshi Airport and by 20 and 22% in the 2020s and the 2050s, respectively at Lyamungu station. They also indicated that more impact is projected to be during 70–80 days after planting and in the entire mid-season (81–140 days) whereby the temperature will be high and the precipitation will be low. In Iran, Emam et al. (2015) projected that the mean annual precipitation is likely to decrease in Razan–Ghahavand basin, while the maximum temperature increases. The changes in these two climate variables resulted in substantial reduction in groundwater recharge, which is the main source of water supply in this area. An increase in the average daily temperature due to climate change will increase reference evapotranspiration resulting in an increased use of irrigation water (Ficklin et al. 2010). Fares et al. (2016) indicated that evapotranspiration responded positively to air temperature rise, as well water requirements for maize and coffee.

3 Implication of Climate Change in Egypt

3.1 *Climate Change and Water Resources*

3.1.1 Precipitation on the Nile Region

Pervious research on the effect of climate change on Nile region and Egypt reported high variation across emission scenarios and employed models, but experts generally agree that the Nile region and Egypt in particular will experience further warming (UNEP 2014). The uncertainty about the increase or decrease in precipitation near the sources of the Nile, as well as variations in temperature could have a larger than expected effect on Nile flows because these two factors are also interrelated leads to moderate to extreme effects (Elsaeed 2012). In addition, Nour El-Din (2013) reported that there are some uncertainty about that the effect of future climate change in East Africa, where roughly two-thirds of the general circulation models project an increase in precipitation on the Blue Nile; where the Blue Nile contributes with more than 75% of the Nile flows. Whereas, one-third of the general circulation models expected reduced precipitation.

Beyene and Meissner (2010) indicated that Nile flows will first increase until mid-century, and then begin declining under scenarios A2 and B1 as a result of evaporation increase and decline in rainfall. On the contrary, a simulation study using HEC-HMS 3.5 model simulate the effects of two climate change scenarios on the hydrological regimes of the upper Blue Nile catchment concluded that Nile river flows would increase from 745 to 127% for these two scenarios (Gebre and Ludwig 2015). Whereas, Adem et al. (2016) used A2 and B2 emission scenarios of the HadCM3 from the Hadley Centre in the UK to study climate change effects using the SWAT model and concluded that both scenarios would lead to an increase in precipitation from September to February and a notable decrease from May to August, where mean annual stream flow would increase by 10% (A2 scenario) or 6% (B2 scenario) by the end of the century.

3.1.2 Precipitation on the North Coast of Egypt

In the north coast of Egypt, where agriculture depends on rain fall, Barbi (2014) indicated that Egypt's precipitation may decrease due to climate change, with an annual

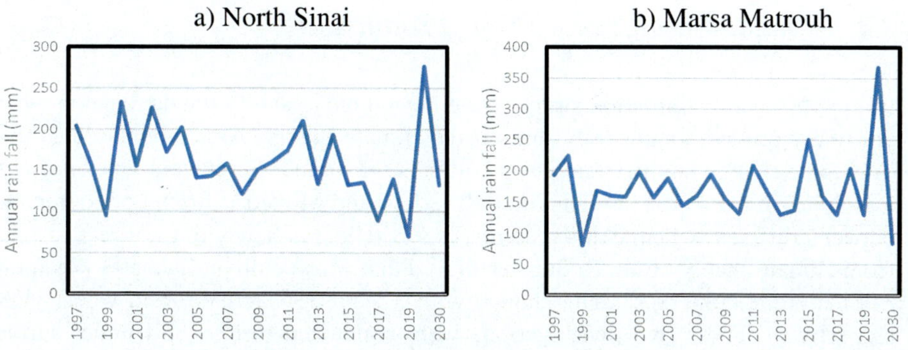

Fig. 1 Comparison between annual amounts of rain fall values from 1997 to 2020 and its projected value in 2030 in **a** North Sinai and **b** Marsa Matrouh

decline up to 5, 8 and 13% by 2030, 2050, and 2100, respectively. Consequently, Ouda et al. (2016a) stated that the projected value of annual rainfall in northeastern coast of Egypt in 2030 was found to be the lowest than its counterpart values from 1997 to 2014, except for the amount of rain fall in 1999, which was lower (Fig. 1a). Rain fall deviation in 2030 from the average of 18 years was −27%, which considered deficit according to the classification of Kumar et al. (2009). Similarly, in the northwestern coast of Egypt, the projected annual rain fall value in 2030 is expected to be the lowest compared to the recorded values from 1997 to 2014 (Fig. 1b). Rainfall deviation in 2030 from the average of 18 years was −95%, which considered scantly according to the classification of Kumar et al. (2009).

3.2 Climate Change and Water Needs for the Cropped Area

Therefore, it is virtually certain that the Nile Basin has become warmer and the warming is expected to continue. Ouda et al. (2016b) reported that using A1B climate change scenario in Egypt to calculate annual evapotranspiration (ETo) values in 2020, 2030 and 2040 and compared it with ETo values in 2014, revealed an increase by 7, 9 and 13% under 2020, 2030 and 2040, respectively. Additionally, Attaher et al. (2006) calculated national irrigation water demand in 2050 and 2100, and they reported an increase by 7 and 11%, respectively.

Sea-level rise in the Nile delta is also expected to occur, which will increase land salinization and water needs for leaching that salts. Therefore, more pressure will be put on water resources in the Nile delta, which is the home of more than 35 million people and providing 63% of Egypt's agricultural production (World Bank 2020) and that will cause many social problems.

3.3 *Climate Change and Crops Production*

In Egypt, several simulation studies have been done to assess the damage on wheat and maize production under the anticipated climate change. Assessment of the effect of climate change on different wheat cultivars grown in different soil types in Egypt was done. In the studies by Khalil et al. (2009) and Moneir (2012), both compared between different wheat cultivars with respect to its resiliency to the stress resulted from climate change. Both studies confirmed that wheat cultivar Sakha93 was found to be more tolerant to climate change with 32 and 35% reduction in wheat yield, respectively. These two investigations were conducted in clay soil under surface irrigation. Whereas, in salt affected soil, the effect of climate change on wheat was higher, namely 45% (Ouda et al. 2012).

Comparison between several wheat cultivars sown in different sowing dates were done in a study conducted in silty clay soil and under sprinkler system by Noreldin et al. (2012). They indicated that Sakha93 planted in 24th of November attained the lowest yield loss under the studied scenario of climate change, namely 36%. Furthermore, Morsy (2015) simulated the effect of climate change on wheat in 2030 and projected 10–12% loss in clay soil under surface irrigation.

Ouda et al. (2010) imitated the farmer traditional practices in growing wheat, where fertilizers were applied by broadcasting on the soil. The study revealed that this practice reduced wheat yield by 30%. In another study by Taha (2012) stated that farmers will lose 38% of wheat yield in 2030, if they continue their practices of irrigation with large amounts of irrigation water and fertilizers under sprinkler irrigation.

Similar analysis was done for surface irrigated maize grown in clay soil, where the tolerance of several maize hybrids were tested by Ouda et al. (2009). They found that maize hybrid TWC324 was found to be more tolerant to climate change stress, where its yield loss were 38%. In silty clay soil and under drip system, Ouda et al. (2012) tested four maize hybrids sown under three sowing dates under climate change and found that the lowest yield loss obtained from hybrid SC128 planted on 2nd week of May, namely 26%. Taha (2012) found that farmer practice in maize management (irrigation and fertilizer) in sandy soil under sprinkler system could reduce its yield by 42%. Furthermore, Morsy (2015) simulated the effect of climate change on maize and projected 13–15% loss in maize yield.

4 Source of Emission of Greenhouse Gases

4.1 *Emission from the Soil*

Gaillard et al. (2016) indicated that soil texture is an important variable in controlling N_2O emission. Soil texture can be classified with respect to its ability to evolve greenhouse gases emission. Coarse-textured soils with higher soil moisture have higher tendency to evolved CO_2 than clay-textured soils (Sun et al. 2014). Whereas, Xie et al. (2013) stated that higher fluxes of CH_4 and N_2O were found in sand-textured than in clay-textured soils, which is induced by the applied soil management and by soil moisture content (Chen et al. 2008). On the other hand, the emission of N_2O in soils of fine texture were greater than that of course-textured soils (Tan et al. 2009).The largest N_2O emissions is found in silt loam soils, compared to loamy sand and sandy loam soils (Gaillard et al. 2016).

Emissions of greenhouse gases from the soils is also controlled by soil pH value. Wang et al. (2010) reported that that CO_2 emission from the soil increased due to addition of N fertilizer to soil under high pH value. Emission of N_2O may increase when soil pH was close to being at neutral value but decreased when pH exceeded the neutral value (Mørkved et al. 2007).

Soil organic carbon or carbon storage is determined by the environmental conditions (climatic factors) among other factors (Bot and Benites 2005). Soil microbial community decomposes the carbon compounds in the soils to extract chemical energy from them, which result in the emission of CO_2. Furthermore, higher temperatures promote the faster breakdown of soil organic matter and the release of CO_2 and CH_4 into the atmosphere through increased soil respiration (Kuzyakov et al. 2019).

4.2 Emission from Rice Fields

Furthermore, under low content of oxygen, namely under waterlogged soils, anaerobic decomposition of organic carbon compounds occurs results in the emission of CH_4 to the atmosphere (Magdoff and Van 2010). Thus, rice cultivation is one of the main sources of global greenhouse gases emissions including CH_4 and N_2O (Zheng et al. 2004). Linquist et al. (2012) reported that a paddy rice field emits higher values of these two gases than a wheat or maize field by 467 and 169%, respectively. Paddy fields contribute about 5–19% of CH_4 emissions to the anthropogenic global CH_4 budget (Smith et al. 2014). Flooding condition in rice field emits a maximum amount of CH_4, in addition to a large amount of N_2O produced in intermittently flooded rice and under transition periods between crops (Zhao et al. 2011). CH_4 is produced by the Methanogenic archaea, and a portion of it is oxidized by the methanotrophic bacteria, whereas the activity of microbial nitrification and denitrification together contribute about 70% of N_2O emission (Braker and Conrad 2011).

N_2O emission greatly increased under continuous inundation and draining cycles from rice cropping systems (Yao et al. 2012). During rice-growing season, N_2O is produced due to alternate wetting/drying period in the underground saturated soil layer (Zheng et al. 2004). Under flooding condition, significant N_2O emission takes place predominately through the rice plants, where rice plants act as a conduit for dissolved gases from the root zone to the atmosphere (Yan et al. 2000).

4.3 *Burning of Crops Residues*

The burning of crop residues contributes in greenhouse gases emissions in the atmosphere, where it increase the rate of CO_2, CH_4 and N_2O and other trace gases (Smith et al. 2014). Tubiello et al. (2014) reported that burning of savanna in Africa represents 25% of the entire agricultural emissions. In Asia, the burning of crop residue is highly practiced, especially rice straw.

In Egypt, the main sources of biomass waste are crop residues, followed by municipal solid waste, animal waste, and sewage waste (Said et al. 2013). It was estimated that about 3.1 Mt/year of rice straw are directly burned in open field (FAO 2013), as well as other crops residues (Abdelhady et al. 2014).

5 Climate-Smart Practices to Reduce Emission of Greenhouse Gases

5.1 *Carbon Sequestration in the Soil*

Soil carbon sequestration implies the removal of atmospheric CO_2, by plants and stored it in soil organic matter (Lal 2004). It have been reported that intensive modern agriculture to produce food usually results in a considerable soil degradation, soil carbon depletion and release CO_2 to the atmosphere (Plaza-Bonilla et al. 2015). To overcome these setbacks, it is very imperative to follow appropriate and climate-smart practices (Lal 2006). No-tillage can increase soil carbon aggregation rapidly, particularly at the soil surface (West and Post 2002). Naveen and Babalad (2018) indicated that the adoption of conservation tillage practices can be helpful in the reductions of emissions of CO_2 into the atmosphere and similarly can be supportive in the sequestration of carbon in the soil. Ariani et al. (2017) reported that zero tillage resulted less emissions than conventional tillage. Pakala and Socolow (2004) reported that the adoption of conservation tillage globally can sequester 25 Gt C over the next 50 years, which can be helpful in the stabilization of atmospheric carbon.

Application of biochar into soil can mitigate the emission of CO_2, CH_4, and N_2O (Li et al. 2018). Furthermore, it was found that organic farming or animal manure application can potentially sequestrate more C to the soils and thus convert the soils to a net CO_2 sink (Gattinger et al. 2012). Diacono et al. (2019) indicated that the cow manure with vegetable wastes and compost with green manure emitted 31.4 and 8.4 kg CO_2 per Mg of compost, respectively, which is lower than carbon emissions of biofertilizers application.

Intercropping is an excellent practice for soil conservation to sustain crop production (Dwivedi et al. 2015). Intercropping legume crops with cereal crops proved to have high benefits to the soil. In addition to its role in biological nitrogen fixation, it has the capacity to store 30% higher soil organic carbon when compared to other species (Kumar et al. 2018). Recent studies suggest greater input of carbon into the soil through root residues in intercropping systems, as compared to sole crops (Li et al. 2011). Cong et al. (2015) indicate that soil carbon sequestration potential of strip intercropping is high, where intercropping can contribute to multiple agro-ecosystem services by increased yield, better soil quality and soil carbon sequestration. It was reported that cover crops increased soil organic carbon stock by 9–10% (Bolinder et al. 2020). It also increase carbon in soils as these materials have high carbon content (Han et al. 2016).

In Egypt, Zohry et al. (2020) stated that an increase in soil organic matter was observed when pea was intercropped with sunflower by 8% and the increase was 9% when pea intercropped with wheat. Abbady et al. (2016) indicated that intercropping pea with sunflower increased carbon sequestration in the soil, compared to the soil cultivated with sole sunflower.

5.2 *Reduction of Greenhouse Gases Emissions from Rice Fields*

Several field managements were studied to reduced CH_4 and N_2O form rice fields. Hadi et al. (2010) indicated that intermittent drainage could reduce greenhouse gases emission by around 14.7 to 68.5% compared to continuous flooding. Whereas, alternate wetting and drying could reduce the emissions by 33–39%, compared to continuous flooding with no yield differences (Setyanto et al. 2017). Intermittent irrigation and saturated soil have less CH_4 emission (around 53–67%), compared to continuous flooding (Husin et al. 1995). Additionally, applications of fresh and decomposed straw could reduce N_2O emission by around 33 and 28% compare to without straw application (Wihardjaka et al. 2010). Water savings techniques applied to rice field can largely contribute to reduction of these two gases (Djaman et al. 2018). Furthermore, conversion from paddy rice cultivars to upland cultivars can contribute in reduced of greenhouse gases (Nishimura et al. 2008).

In Egypt, rice belt is located in the Northern Nile Delta, where CH_4 emission is representing about 53.3% of the agricultural greenhouse gases emissions (Farag et al. 2013a). Recently, the total CH_4 emissions from rice cultivation in Egypt are steady decreasing due to switching from long-season traditional rice cultivars to early-maturing short-duration cultivars (Farag et al. 2013b). Furthermore, the cultivation of new upland rice cultivars is now expanding in Egypt.

5.3 *Biogas Production*

In the rural areas, there are huge amounts of crops residual available for production of biogas (Zayat et al. 2015), otherwise it will be burn releasing greenhouse gases. Biogas is produced from anaerobic degradation of organic materials. It consists mainly of CH_4 and CO_2, and traces of hydrogen sulfide and mercaptanes, as well as residual amounts of oxygen and nitrogen (Cesaro and Belgiorno 2015). Harold (2007) indicated that the concentration of CH_4 and CO_2 in biogas varies from 55 to 65% and 40 to 45%, respectively. Biogas is generated from an organic anaerobic digestion process, which requires a symbiotic mixture of certain bacteria and organic material (Christopher et al. 2010).

Biogas units in the households could use crops residual and animal manure, which will be fermented in biogas digesters and a significant amount of methane is resulted that could be used in cooking, heating, and generating electricity (Zayat et al. 2015). Furthermore, the potential utilization of the digestate as fertilizers can also reduce dependence on mineral fertilizers (Pöschl and Owende 2010), which also reduce emission of greenhouse gases. Thus, biogas generation serves three important functions: waste removal, environmental management, and energy production (Ford 2007).

6 Conclusion

Climate-smart practices can highly participate in the reduction of greenhouse gases emission from agricultural fields. The main source of greenhouse gases are emission from the soil, which can reduced by practices encourage carbon sequestration in the soil, such as conservation agriculture, and implementing intercropping systems with legumes. Another important source of greenhouse gases is emission from rice fields, where conversion from long season rice cultivars to short season cultivars and from paddy rice cultivars to upland rice cultivars can contribute in its reduction. The third important contributor in greenhouse gases emission is the burning of crops residues. This source of emission can be eliminated if these wastes were used in biogas production.

References

Abbady K, El-Maaz E, Ahmed H, Zohry AA (2016) Carbon sequestration as a function of intercropping management practice and different nitrogenous fertilizer types. J Soil Sci Agric Eng 7(8):565–586

Abbaspour KC, Faramarzi M, Ghasemi SS, Yang H (2009) Assessing the impact of climate change on water resources in Iran. Water Resour Res 45(10):1–16

Abdelhady S, Borellob D, Shabanb A, Rispolib F (2014) Viability study of biomass power plant fired with rice straw in Egypt. Energy Proc 61:211–215. https://doi.org/10.1016/j.egypro.2014.11.1072

Adem A, Tilahun SA, Ayana EK, Worqlul AW, Assefa TT, Dessu SB, Melesse AM (2016) Climate change impact on sediment yield in the Upper Gilgel Abay catchment, Blue Nile Basin, Ethiopia. In: Melesse AM, Abtew W (eds) Landscape dynamics, soils and hydrological processes in varied climates. Springer International Publishing, pp 615–642

Adhikari U, Nejadhashemi AP, Woznicki SA (2015) Climate change and eastern Africa: a review of impact on major crops. Food Energy Secur 4:110–132
Ali R, McFarlane D, Varma S, Dawes W, Emelyanova I, Hodgson G, Charles S (2012) Potential climate change impacts on groundwater resources of south-western Australia. J Hydrol 475:456–472
Allison SD, Treseder KK (2011) Climate change feedbacks to microbial decomposition in boreal soils. Fungal Ecol 4:362–374
Ariani M, Yulianingrum H, Setyanto P (2017) Emisi Gas Rumah Kaca dan Hasil Padi dari Cara Olah Tanah dan Pemberian Herbisida Di Lahan Sawah MK 2015 Jurnal Ilmu Lingkungan 15:74–82
Asseng S, Ewert F, Rosenzweig C, Jones JW, Hatfield JL (2013) Uncertainty in simulating wheat yields under climate change. Nat Clim Change 3:827–832
Attaher S, Medany M, AbdelAziz AA, El-Gendi A (2006) Irrigation-water demands under current and future climate conditions in Egypt. In: The 14th annual conference of the Misr Society of Agricultural Engineering, pp 1051–1063
Banziger M, Diallo AO (2004) Progress in developing drought and N stress tolerant maize cultivars for Eastern and Southern Africa. In: Integrated approaches to higher maize productivity in the New Millennium. Proceedings of the 7th Eastern and Southern Africa regional maize conference, CIMMYT/KARI; Nairobi, Kenya, pp 189–194
Bates BC, Kundzewicz ZW, Wu S, Palutikof JP (eds) (2008) Climate change and water. technical paper of the intergovernmental panel on climate change, IPCC Secretariat, Geneva, p 210
Barbi V (2014) Adaptation: key concepts, strategies, and practices. In: International centre for climate governance. www.iccgov.org/wp-content/uploads/2014/.../2014-05-22_Barbi.pdf
Beyene EG, Meissner B (2010) Spatio-temporal analyses of correlation between NOAA satellite RFE and weather stations' rainfall record in Ethiopia. Int J Appl Earth Obs Geoinf 12:69–75
Bolinder MA, Crotty F, Elsen A (2020) The effect of crop residues, cover crops, manures and nitrogen fertilization on soil organic carbon changes in agroecosystems: a synthesis of reviews. Mitig Adapt Strateg Glob Change 25:929–952. https://doi.org/10.1007/s11027-020-09916-3
Bot A, Benites J (2005) The importance of soil organic matter. FAO, Rome. Italy 2005:94
Braker G, Conrad R (2011) Diversity, structure, and size of N_2O–producing microbial communities in soils–what matters for their functioning? Adv Appl Microbiol 75:33–37
Calleja-Cabrera J, Boter M, Oñate-Sánchez L, Pernas M (2020) Root growth adaptation to climate change in crops. Front Plant Sci 11, Article 544. https://doi.org/10.3389/fpls.2020.00544
Cesaro A, Belgiorno V (2015) Combined biogas and bioethanol production: opportunities and challenges for industrial application. Energies 8:8121–8144
Challinor AJ, Simelton ES, Fraser EDG, Hemming D, Collins M (2010) Increased crop failure due to climate change: assessing adaptation options using models and socio-economic data for wheat in China. Environ Res Lett 5:034012
Chanda A, Maghrawy H, Sayour H, Gummadidala PM, Gomaa OM (2020) Impact of climate change on plant-associated fungi. In: Omran E-S, Negm AM (eds) Climate change impacts on agriculture and food security in Egypt: land and water resources—smart farming—livestock, fishery, and aquaculture. Springer Nature Switzerland AG
Chen ST, Huang Y, Zou JW (2008) Relationship between nitrous oxide emission and winter wheat production. Biol Fertil Soil (berlin) 44(7):985–989
Christopher IT, Hassan MG, Ali SF (2010) Integrated portable biogas systems for managing organic wastes. In: Proceedings of WSEAS conference, Tunisia
Cong W-F, Hoffland E, Li L, Six J, Sun J-H, Bao X-G, Zhang F-S, Der Werf WV (2015) Intercropping enhances soil carbon and nitrogen. Glob Change Biol 21(4):1715–1726. https://doi.org/10.1111/gcb.12738
De Costa WAJM (2011) A review of the possible impacts of climate change on forests in the humid tropics. J Natl Sci Found Sri 39:281–302. https://doi.org/10.4038/jnsfsr.v39i4.3879
Diacono M, Persiani A, Testani E, Montemurro F, Ciaccia C (2019) Recycling agricultural wastes and by-products in organic farming: biofertilizer production, yield performance and carbon footprint analysis. Sustainability 11:3824. https://doi.org/10.3390/su11143824

Djaman K, Mel VC, Diop L, Sow A, El-Namaky R, Manneh B (2018) Effects of alternate wetting and drying irrigation regime and nitrogen fertilizer on yield and nitrogen use efficiency of irrigated rice in the Sahel. Water 10(711):20–25

Dwivedi A, Dev I, Kumar V, Yadav RS, Yadav M, Gupta D, Singh A (2015) Potential role of maize-legume intercropping systems to improve soil fertility status under Smallholder farming systems for sustainable agriculture in India. Tomar1 Inter J Life Sci Biotechnol Pharma Res 4(3):215–222

Elsaeed G (2012) Effects of climate change on Egypt's water supply. In: Fernando HJS et al (eds) National security and human health implications of climate change. NATO science for peace and security series C: environmental security. © Springer Science+Business Media B.V. https://doi.org/10.1007/978-94-007-2430-3_30

Emam AR, Kappas M, Hosseini SZ (2015) Assessing the impact of climate change on water resources, crop production and land degradation in a semi-arid river basin. Hydrol Res 46(6):854–870

Eyshi Rezaei E, Siebert S, Hüging H, Ewert F (2018) Climate change effect on wheat phenology depends on cultivar change. Sci Rep 2018(8):4891. https://doi.org/10.1038/s41598-018-23101-2

FAO (2013) Rice market monitor, Nov 2013

Farag AA, Radwan HA, Abdrabbo MAA, Heggi MAM (2013a) Inventory of the greenhouse gas emissions from rice in the Nile Delta by using emission models. Egypt. J Agric Res. 91(2b):917–937

Farag AA, Radwan HA, Abdrabbo MAA, Heggi MAM, McCarl BA (2013b) Carbon footprint for paddy rice production in Egypt Nat Sci 11(12):36–45. http://www.sciencepub.net/nature

Fares A, Awal R, Fares S, Johnson AB (2016) Irrigation water requirements for seed corn and coffee under potential climate change scenarios and Hector Valenzuela. J Water Clim Change 07(1)3:9–51

Faurès JM, Bartley D, Bazza M, Burke J, Hoogeveen J, Soto D, Steduto P (2013) Climate smart agriculture sourcebook. FAO, Rome

Ficklin DL, Luo Y, Luedeling E, Zhang M (2010) Climate change sensitivity assessment of a highly agricultural watershed using SWAT. J Hydrol 374:16–29

Ford S (2007) Advances in biogas. Pira International Ltd, Leatherhead, pp 13–14

Gaillard R, Duval BD, Osterholz WR, Kucharik CJ (2016) Simulated effects of soil texture on nitrous oxide emission factors from corn and soybean agroecosystems in Wisconsin. J Environ Qual 45:1540–1548. https://doi.org/10.2134/jeq2016.03.0112

Gattinger A, Muller A, Haeni M, Skinner C, Fliessbach A, Buchmann N, Mäder P, Stolze M, Smith P, Scialabba Nel-H, Niggli U (2012) Enhanced top soil carbon stocks under organic farming. Proc Natl Acad Sci USA 109:18226–18231

Gebre SL, Ludwig F (2015) Hydrological response to climate change of the Upper Blue Nile River Basin: based on IPCC fifth assessment report (AR5). J Climatol Weather Forecast 3:121. https://doi.org/10.4172/2332-2594.1000121

Girones R, Ferrús MA, Alonso JL, Rodriguez-Manzano J, Calgua B, de Abreu CA, Hundesa A, Carratala A, Bofill-Mas S (2010) Molecular detection of pathogens in water—the pros and cons of molecular techniques. Water Res 44:4325–4339. https://doi.org/10.1016/j.watres.2010.06.030

Hadi A, Inubushi K, Yagi K (2010) Effect of water management on greenhouse gas emissions and microbial properties from paddy fields of Indonesia and Japan. Paddy Water Environ 8:319–324

Han P, Zhang W, Wang G, Sun W, Huang Y (2016) Changes in soil organic carbon in croplands subjected to fertilizer management. A global meta-analysis. Sci Rep 6:27199. https://doi.org/10.1038/srep27199

Harold H (2007) Alternative energy sources-biogas production. In: The proceedings of London Swine conference

Holmén H (2003) Reflections on natural reconditions for a green revolution in Africa. Department of Geography, Linkoping University, The Tema Institute, SE-581 83, Linkoping

Huang BR, Taylor HM, McMichael BL (1991) Effects of temperature on the development of metaxylem in primary wheat roots and its hydraulic consequence. Ann Bot 67:163–166. https://doi.org/10.1093/oxfordjournals.aob.a088115

Hund A, Fracheboud Y, Soldati A, Stamp P (2008) Cold tolerance of maize seedlings as determined by root morphology and photosynthetic traits. Eur J Agron 28:178–185. https://doi.org/10.1016/j.eja.2007.07.003

Husin YA, Murdiyarso D, Khalil MAK, Rasmussen RA, Shearer MJ, Sabiham S, Sunar A, Adijuwana H (1995) Methane flux from Indonesian wetland rice: the effects of water management and rice variety. Chemosphere 31:3153–3180

IPCC (2007) Contribution of working group II to the fourth assessment report of the intergovernmental panel on climate variability. Cambridge University Press, Cambridge, p 60

IPCC (2014) Climate change 2014: impacts, adaptation, and vulnerability. Part A: Global and sectoral aspects. Summary for policymakers, pp 1–32. Contribution of working group II to the fifth assessment report of the intergovernmental panel on climate change (IPCC). Cambridge University Press, Cambridge and New York, NY

IPCC (2018) Summary for policymakers of IPCC special report on global warming of 1.5_C approved by governments—IPCC. Available at Ipcc.ch. https://www.ipcc.ch/2018/10/08/summary-for-policymakers-of-ipcc-special-report-on-lobal-warming-of-1-5c-approved-bygovernments/

Joshi M, Fogelman E, Belausov E, Ginzberg I (2016) Potato root system development and factors that determine its architecture. J Plant Physiol 205:113–123. https://doi.org/10.1016/j.jplph.2016.08.014

Khalil FA, Farag H, El Afandi G, Ouda SA (2009) Vulnerability and adaptation of wheat to climate change in Middle Egypt. In: Proceeding of the 13th international conference on water technology, Egypt, 12–15 March

Knox J, Daccache A, Hess T, Haro D (2016) Meta-analysis of climate impacts and uncertainty on crop yields in Europe. Environ Res Lett 11:113004

Kumar S, Meena R, Lal R, Yadav GS, Mitran T, Meena BL, Dotaniya ML, EL-Sabagh A (2018) Role of legumes in soil carbon sequestration. In: Meena RS et al (eds) Legumes for soil health and sustainable management. Springer Nature Singapore Pte Ltd. https://doi.org/10.1007/978-981-13-0253-4_4

Kumar NM, Murthy CS, SeshaSai MVR, Roy PS (2009) On the use of Standardized Precipitation Index (SPI) for drought intensity assessment. Meteorol Appl. https://doi.org/10.1002/met.136

Kuzyakov Y, Horwath WR, Dorodnikov M, Blagodatskaya E (2019) Review and synthesis of the effects of elevated atmospheric CO_2 on soil processes: no changes in pools, but increased fluxes and accelerated cycles. Soil Biol Biochem 128:66–78

Lal R (2004) Soil carbon sequestration to mitigate climate change. Geoderma 123:1–22

Lal R (2006) Enhancing crop yields in the developing countries through restoration of soil organic carbon pool in agricultural lands. Land Degrad Dev 2(17):197–209. https://doi.org/10.1002/ldr-696

Lal R (2015) Sequestering carbon and increasing productivity by conservation agriculture. J Soil Water Conserv 70(3):55A-62A. https://doi.org/10.2489/jswc.70.3.55A

Li L, Sun JH, Zhang FS (2011) Intercropping with wheat leads to greater root weight density and larger below-ground space of irrigated maize at late growth stages. Soil Sci Plant Nutr 57:61–67

Li Y, Hu S, Chen J, Müller K, Li Y, Fu W, Lin Z, Wang H (2018) Effects of biochar application in forest ecosystems on soil properties and greenhouse gas emissions: a review. J Soils Sediments 18:546–563

Linquist B, Van Groenigen KJ, Adviento-Borbe MA, Pittelkow C, Van Kessel C (2012) An agronomic assessment of greenhouse gas emissions from major cereal crops. Global Change Biol 18:194–209

Liu J, Yang H (2010) Spatially explicit assessment of global consumptive water uses in cropland: green and blue water. J Hydrol 384:187–197

Liwenga ET, Kangalawe RYM, Lyimo JG, Majule AE, Ngana JO (2007) Research protocols for assessing the impact of CC & V in rural Tanzania: water, food systems, vulnerability and adaptation. START/PACOM

Magdoff F, Van EH (2010) Building soils for better crops, 3rd edn. SARE Outreach Publications, Brentwood, CA, USA

Mbow C, Rosenzweig C, Barioni LG, Benton TG, Herrero M, Krishnapillai M, Liwenga E, Pradhan P, Rivera-Ferre MG, Sapkota T, Tubiello FN, Xu Y (2019) Food security. In: Shukla PR, Skea J, Calvo Buendia E, Masson-Delmotte V, Pörtner H-O, Roberts DC, Zhai P, Slade R, Connors S, van Diemen R, Ferrat M, Haughey E, Luz S, Neogi S, Pathak M, Petzold J, Portugal Pereira J, Vyas P, Huntley E, Kissick K, Belkacemi M, Malley J (eds) Climate change and land: an IPCC special report on climate change, desertification, land degradation, sustainable land management, food security, and greenhouse gas fluxes in terrestrial ecosystems

Meadows ME, Hoffman TM (2003) Land degradation and climate change in South Africa. Geogr J 169(2):168–177

Mo X-G, Guo R-P, Liu S-X et al (2013) Impacts of climate change on crop evapotranspiration with ensemble GCM projections in the North China Plain. Clim Change 120:299–312

Moneir I (2012) Simulation of the effect of adaptation strategies on improving yield of some crops grown under expected climate change conditions. MSc thesis, Institute of Environmental Research Studies, Ain Shams University, Egypt

Morales D, Rodríguez P, Dell'Amico J, Nicolás E, Torrecillas A, Sánchez-Blanco MJ (2003) High-temperature preconditioning and thermal shock imposition affects water relations, gas exchange and root hydraulic conductivity in tomato. Biol Plant 47:203–208. https://doi.org/10.1023/B:BIOP.0000022252.70836.fc

Mørkved PT, Dörsch P, Bakken LR (2007) The N_2O product ratio of nitrification and its dependence on long-term changes in soil pH. Soil Biol Biochem 39:2048–2057. https://doi.org/10.1016/j.soilbio.2007.03.006

Morsy M (2015) Use of regional climate and crop simulation models to predict wheat and maize productivity and their adaptation under climate change. PhD thesis, Faculty of Science Al-Azhar University

Msowoya K, Madani K, Davtalab R, Mirchi A, Lund JR (2016) Climate change impacts on maize production in the warm heart of Africa. Water Resour Manag 30:5299–5312. https://doi.org/10.1007/s11269-016-1487-3

Nagel KA, Kastenholz B, Jahnke S, van Dusschoten D, Aach T, Mühlich M et al (2009) Temperature responses of roots: impact on growth, root system architecture and implications for phenotyping. Funct Plant Biol 36:947. https://doi.org/10.1071/FP09184

Naveen KBT, Babalad HB (2018) Soil organic carbon, carbon sequestration, soil microbial biomass carbon and nitrogen and soil enzymatic activity as influenced by conservation agriculture in pigeonpea and soybean intercropping system. Int J Curr Microbiol App Sci 7(03):323–333. https://doi.org/10.20546/ijcmas.2018.703.038

Nelson D, Adger W, Brown K (2007) Adaptation to environmental change: contributions of a resilience framework. Ann Rev Environ Resour 32:395–419

Noreldin T, Abdrabbo M, Ouda S (2012) Increasing water productivity for wheat grown under climate change conditions. In: 10th international conference of Egyptian Soil Science Society (ESSS) and 4th international conference of Water Requirements & Metrology Dept. 5–7 November, Ameria, Egypt (Cl 06)

Nour El-Din MM (2013) Climate change risk management in Egypt: proposed climate change adaptation strategy. Ministry of Water Resources & Irrigation. Prepared for UNESCO-Cairo Office

Nishimura S, Yonemura S, Sawamoto T, Shirato Y, Akiyama H, Sudo S, Yagi K (2008) Effect of land use change from paddy rice cultivation to upland crop cultivation on soil carbon budge of a cropland in Japan. Agric Ecosyst Environ 125:9–20

Ouda SA, Khalil FA, Yousef H (2009) Using adaptation strategies to increase water use efficiency for maize under climate change conditions. In: Proceeding of 13th international conference on water technology, Egypt

Ouda SA, Sayed M, El Afandi G, Khalil FA (2010) Developing an adaptation strategy to reduce climate change risks on wheat grown in sandy soil in Egypt. In: Proceeding of 10th international conference on development of dry lands, Egypt

Ouda S, Abdrabbo M, Noreldin T (2012) Effect of changing sowing dates and irrigation scheduling on maize yield grown under climate change conditions. In: 10th international conference of Egyptian Soil Science Society (ESSS) and 4th international conference of Water Requirements & Metrology Dept. 5–7 November, Ameria, Egypt (Cl 07)

Ouda S, Ewise M, Noreldin T (2016a) Projection of productivity of cultivated crops in rain-fed areas in Egypt under climate change. Cogent Food Agric 2(1):1136256

Ouda S, Noreldin T, Hosny M (2016b) Evapotranspiration under changing climate. In: Major Crops and Water Scarcity in Egypt. Springer Publishing House, pp 1–22. ISBN: 978-3-319-21770

Pakala S, Socolow R (2004) Stabilization wedges: solving the climate problem for the next 50 years with current technologies. Science 305:968–972

Parry M, Rosenzweig C, Inglesias A, Livermore M, Gischer G (2004) Effects of climate change on global food production under SRES emissions and socio-economic scenarios. Glob Environ Change 14:53–67

Paymard P, Bannayan M, Haghighi RS (2018) Analysis of the climate change effect on wheat production systems and investigate the potential of management strategies. Nat Hazards 91:1237. https://doi.org/10.1007/s11069-018-3180-8

Plaza-Bonilla D, José Luis A, Cantero-Martinez C, Fanlo R, Iglesias A, Alvaro-Fuentes J (2015) Carbon management in dry land agricultural systems. A review. Agron Sustain Dev 35(4):1319–1334. https://doi.org/10.10007/s13593-015-0326-x

Porter JR, Xie L, Challinor AJ, Cochrane K, Howden M, Iqbal MM, Lobell DB, Travasso MI (2014) Food security and food production systems. Climate change 2014: impacts, adaptation and vulnerability. Working group II contribution to the IPCC 5th assessment report, Geneva, Switzerland

Pöschl MS, Ward P Owende (2010) Evaluation of energy efficiency of various biogas production and utilization pathways. J Appl Sci87:3305–3321

Ravindranath NH, Manuvie R, Lakshmi CS (2010) Biofuels and climate change. In: Rosillo-Calle F, Johnson FX (eds) Food versus fuel. Zed Books Ltd, London and New York

Rinaldi M (2009) A simulation approach to investigate options for mitigation and crop adaption to climate change in Mediterranean area. In: 2009 IOP conference series: earth environ science. https://doi.org/10.1088/17551307/6/7/372038

Rotich SC, Mulungu DMM (2017) Adaptation to climate change impacts on maize water requirements. J Water Clim Change 08(2):274–292

Said N, EL-Shatoury SA, Diaz LF, Zamorano M (2013) Quantitative appraisal of biomass resources and their energy potential in Egypt. Renew Sustain Energy Rev 24:84–91

Schiermeier Q (2014) The parched planet: water on tap. Nature 510:326–328. https://doi.org/10.1038/510326a

Setyanto P, Pramono A, Adriany TA, Susilawati HL, Tokida T, Padre AT, Minamikawa K (2017) Alternate wetting and drying reduces methane emission from a rice paddy in Central Java, Indonesia without yield loss Soil Sci. Plant Nutr 64:23–30

Shiferaw B, Prasanna BM, Hellin J, Bänziger M (2011) Crops that feed the world 6. Past successes and future challenges to the role played by maize in global food security. Food Secur 307–327

Singh R, Khanna V (2021) Climate change and food systems: implications on food security. In: Hebsale Mallappa VK, Shirur M (eds) Climate change and resilient food systems. Springer Nature Singapore Pte Ltd. https://doi.org/10.1007/978-981-33-4538-6_3

Smith JB, McCarl BA, Kirshen P, Jones R, Deck L, Abdrabo MA, Borhan M, El-Ganzori A, El-Shamy M, Hassan M, El-Shinnawy I, Abrabou M, Hassanein MK, El-Agizy M, Bayoumi M, Hynninen R (2014) Egypt's economic vulnerability to climate change. Int Res Clim Res. 62:59–70. https://doi.org/10.3354/cr01257

Sommer R, Glazirina M, Yuldashev T, Otarov A, Ibraeva M, Martynova L (2013) Impact of climate change on wheat productivity in Central Asia. Agric Ecosyst Environ 178:78–99

Sun L, Li L, Chen Z, Wang J, Xiong Z (2014) Combined effects of nitrogen deposition and biochar application on emissions of N_2O, CO_2 and NH_3 from agricultural and forest soils. Soil Sci Plant Nutr 60:254–265. https://doi.org/10.1080/00380768.2014.885386

Tadross M, Wolski P (2010) Climate change modeling for the Pangani Basin to support integrated water management planning process (Final report). Climate Systems Analysis Group, SA

Taha A (2012) Effect of climate change on maize and wheat grown under fertigation treatments in newly reclaimed soil. PhD thesis, Tanta University, Egypt

Tan IYS, van Esa HM, Duxburya JM, Melkoniana JJ, Schindelbecka RR, Larry D, Dean-Hivelya GW, Moebiusa BN (2009) Single-event nitrous oxide losses under maize production as affected by soil type, tillage, rotation, and fertilization. Soil till Res (amsterdam). 102(1):19–26. https://doi.org/10.1016/j.still.2008.06.005

Teixeira EI, Fischer G, Velthuizen HV, Walter C, Ewert F (2013) Global hot spots of heat stress on agricultural crops due to climate change. Agric for Meteorol 170:206–215. https://doi.org/10.1016/j.agrformet.2011.09.002

Tesfaye K, Gbegbelegbe S, Cairns JE, Shiferaw B, Prasanna BM, Sonder K et al (2015) Maize systems under climate change in sub-Saharan Africa: potential impacts on production and food security. Int J Clim Change Strat Manag 247–271

Tubiello FN, Salvatore M, Condor Golec RD, Ferrara AF, Rossi S, Biancalani R, Federici S, Jacobs H, Flammini A (2014) Agriculture, forestry and other land use emissions by sources and removals by Sinks. ESS working paper no. 2. Food and Agriculture Organization of the United Nations, Rome

UNEP (2014) Green economy scoping study: Egypt. https://www.greengrowthknowledge.org/sites/default/files/downloads/resource/Green_Economy_Scoping_Study_Egypt_UNEP.pdf

Urban DW, Sheffield J, Lobell DB (2015) The impacts of future climate and carbon dioxide changes on the average and variability of US maize yields under two emission scenarios. Environ Res Lett 10:045003

Vermeulen SJ, Campbell BM, Ingram JSI (2012) Climate change and food systems. Annu Rev Env Resour 37:195–222. https://doi.org/10.1146/annurev-environ-020411-130608

Wang L, Han Z, Zhang X (2010) Effects of soil pH on CO_2 emission from long-term fertilized black soils in Northeastern China. In: Proceedings of conference on environmental pollution and public health (CEPPH 2010). Wuhan, China. Water Conserv 70(3):55A–62A. https://doi.org/10.2489/jswc.70.3.55A

West TO, Post WM (2002) Soil organic carbon by tillage and crop rotation: a global data analysis. Soil Sci Soc Am J 66(6):1930–1946

Wihardjaka A, Tandjung SD, Sunarminto BH, Sugiharto E (2010) Emisi Gas Dinitrogen Oksida Pada Padi Gogorancah Oleh Pemberian Jerami Padi Dan Bahan Penghambat Nitrifikasi Penelitian Pertanian Tanaman Pangan 29:144–151

World Bank (2020) Modernizing irrigation improved water security for farmers in Egypt. https://www.worldbank.org/en/results/2020/04/01/modernizing-irrigation-improved-water-security-for-farmers-in-egypt

World Resource Institute (2000) World GHG emissions flow chart. World Resource Institute, Climate Analysis Indicators Tool (CIAT). http://cait.wri.org/figures.php?page¼/World-FlowChart

Wu C, Huang G, Yu H, Chen Z, Ma J (2014) Impact of climate change on reservoir flood control in the upstream area of the Beijiang river basin, South China. J Hydrometeorol 15(6):2203–2218. http://www.jstor.org/stable/24914895

Xie Z, Xu Y, Liu G, Liu Q, Zhu J, Tu C, Amonette JE, Cadisch G, Yong JWH, Hu S (2013). Impact of biochar application on nitrogen nutrition of rice, greenhouse-gas emissions and soil organic carbon dynamics in two paddy soils of China. Plant Soil 370:527–540. https://doi.org/10.1007/s11104-013-1636-x

Xing-Guo MO, Hu S, Lin Z-H, Liu S-X; Xia J (2017) Impacts of climate change on agricultural water resources and adaptation on the North China Plain. Adv Clim Change Res 8:93e98

Yan W, Hunt LA, Sheng Q, Szlavnies Z (2000) Cultivar evaluation and mega-environment investigation based on GGEbiplot. Crop Sci 40:597−605

Yao Z, Zheng X, Dong H, Wang R, Mei B, Zhu J (2012) A 3–year record of N2O and CH4 emissions from a sandy loam paddy during rice seasons as affected by different nitrogen application rates. Agric Ecosyst Environ 152:1–9

Zayat ME, Hassan MG, Cl T, Haggar SE (2015) Feasibility of biogas utilization in developing countries: Egypt a case study. Austin Chem Eng. 2(2):1017

Zhao C, Piao S, Wang X et al (2017a) Plausible rice yield losses under future climate warming. Nat Plants 3:16202. https://doi.org/10.1038/nplants.2016.202

Zhao CB, Liu SP, Wang X, Lobell DB, Huang Y, Huang M, Yao Y, Bassu S, Ciais P, Durand J-L, Elliott J, Ewert F, Janssens IA, Li T, Lin E, Liu Q, Martre P, Müller C, Peng S, Peñuelas J, Ruane AC, Wallach D, Wang T, Wu D, Liu Z, Zhu Y, Zhu Z, Asseng S (2017b) Temperature increase reduces global yields of major crops in four independent estimates. PNAS 114(35):9326–9331. https://doi.org/10.1073/pnas.1701762114

Zhao X, Min J, Wang S, Shi W, Xing G (2011) Further understanding of nitrous oxide emission from paddy fields under rice/wheat rotation in south China. J Geophys Res Biogeosci 116(G2). https://doi.org/10.1029/2010JG001528

Zheng X, Han S, Huang Y, Wang Y, Wang M (2004) Re-quantifying the emission factors based on field measurements and estimating the direct N_2O emission from Chinese croplands. Glob Biogeochem Cycle 18:GB2018

Zohry A, Ouda S, Abdel-Wahab T (2020) Sustainable intensive cropping to reduce irrigation-induced erosion: Intercropping systems under surface irrigation practice. Moroccan J Agric Sci 1(2):63–71